T0270866

Nevill Mott

NEVILL MOTT

Reminiscences and Appreciations

Edited by

E. A. Davis

University of Leicester

CRC Press
Taylor & Francis Group
Boca Raton London New York

CRC Press is an imprint of the
Taylor & Francis Group, an **informa** business

Contents

Contents

Contents

Education

Taylor & Francis

Late Cambridge

Contents

Contents

Religion

Preface

I am very pleased to contribute in a small way by writing this preface as Managing Director of Taylor & Francis. This book is written and published in honour and memory of Professor Sir Nevill Mott and as a tribute to his tremendous work during his life. The idea was proposed by Professor E. A. (Ted) Davis of the University of Leicester, who was a collaborator with Nevill, at the Cavendish Laboratory, Cambridge. Ted also compiled and edited the volume.

Having been an employee of Taylor & Francis for over thirty-five years, I clearly remember Nevill's continuous support for the company and his influence on its development. It is with particular gratitude that this preface is written.

He was a true friend of Taylor & Francis and a person who always had a kind word for the employees from the top to the bottom. I remember my first encounter with Nevill, having only joined the company a short while earlier. I passed him in the corridor and he asked me my name and what I did. I told him and he replied with vigour that he was sure I would enjoy my time here. As usual, he emphasized the company's tremendous history and reputation, suggesting that it would provide a good career for me in publishing. He was right, and last year I was very proud to be responsible for organizing Nevill's ninetieth birthday celebrations at Gonville and Caius College, Cambridge.

There are many reminiscenses of people who met and worked with Nevill. The contributions made by all his friends and colleagues provide the reader with an insight into his character. Taylor & Francis wishes to thank all authors for their contributions and is pleased to be able to publish this book in Nevill's memory.

Tony Selvey

Introduction

Edward A. Davis

Professor Sir NEVILL FRANCIS MOTT
CH, KB, ScD, FRS, Chevalier Ordre Nat. du Mérite
Nobel laureate in physics
30 September 1905 – 8 August 1996

The death of Sir Nevill Mott at age 90 brought to a close the illustrious 'life in science' of a man who stood academically, as well as literally, head and shoulders above his contemporaries, many of whom were awed and inspired by his remarkable powers of intuition, his penetrating insight and his versatile creativity.

The staggering length of his scientific career – his first paper was published in 1927 during the early days of quantum theory and his last on high-temperature superconductivity in 1996, four months after his death – must be unprecedented, and the range of his contributions to physics possibly unequalled. He was the author of at least fifteen books and over three hundred scientific papers, covering a range of topics as diverse as wave mechanics, nuclear physics, the properties of metals, ionic crystals, semiconductors and glasses, and the nature of superconductivity.

Nevill Mott was a theoretical physicist, a label far too narrow and restrictive to describe the extent of his activities and the sphere of his influence. His principal posts were at Bristol and Cambridge but he was the 'father' of a much larger community, communicating (without the benefit of e.mail) with hundreds of scientists, theoreticians and experimentalists alike, in numerous countries – via handwritten letters most of which contained gems of ideas that were treasured by their recipients. Unlike many of us, he didn't have a cluttered desk with a massive pile in the in-tray; he worked in 'real time', writing first drafts of papers as ideas occurred to him and answering letters, responding to requests for information, and writing referee's reports, references, etc., often by return of post. His mind was his principal filing cabinet. In these ways he drew together ideas and people from different disciplines in a unique manner.

Early Years

It was natural that Mott should take an early interest in physics. His mother and father both worked at the Cavendish Laboratory at Cambridge University with J. J. Thomson, a few years after JJ's 'discovery' of the electron in

Cavendish research students of 1904: (front row, left to right) C. F. Mott (Nevill's father), H. A. Wilson, Miss Reynolds (Nevill's mother), Professor J. J. Thomson, Miss Slater, P. V. Bevan, O. W. Richardson; (middle row) W. Mackower, A. Wood, S. C. Laws, R. K. McClung, F. Horton, H. L. Cooke, N. R. Campbell, J. H. Field; (back row) J. Satterly, G. Jaffé, R. Hosking, T. B. Vinycomb, G. Owen, P. S. Barlow; (seated on the ground) E. P. Harrison and S. A. Edmonds.

1897. In his autobiography, *A Life in Science*, he records that quite early in his life his parents communicated to him the excitement and importance of physics. His mother taught him at home until he was 10, when he became a weekly boarder at a small preparatory school near Stafford. Before he left that school at age 13 he was, according to his own words, advanced in algebra and Latin and had been introduced to calculus. He spent the next five years as a boarder at Bristol's Clifton College, where a gifted teacher, H. C. Beaven, instilled in him the beauty of mathematics. Later he recalled the excitement he felt when, at about age 16, he realised why any number raised to the power nought equalled unity! In December 1923 he was awarded a major scholarship at St John's College, Cambridge, where he read for the Mathematical Tripos.

In spite of his mathematical ability, Mott was never tempted to be a pure mathematician; physics was his main interest. His first research was undertaken when he was an undergraduate. In the absence of any real suggestions from the head of theoretical physics at the Cavendish, Ralph Fowler, he set about finding his own problem, which he decided should be on the application of wave mechanics to the scattering of charged particles – an active area of experimental investigations led by Ernest Rutherford. This work was published in the *Proceedings of the Royal Society*.

One year after he graduated with first-class honours in 1928, Fowler arranged a grant to enable him to spend a year abroad, which he divided between visits to Copenhagen and Göttingen. At Copenhagen he discussed with Niels Bohr the startling prediction of Dirac in Cambridge that the electron carried a spin and he began work on a related problem. He clearly found it exciting to be in the company of Bohr and other famous scientists of the era; Pauli, Gamow, Heisenberg and Hartree all visited Bohr's famous institute. Before moving on to Göttingen, he returned for a few months to Cambridge, where he solved an important problem relating to the scattering of alpha particles. Chadwick, Rutherford's second in command, was clearly impressed and took Mott along to see Rutherford, who exclaimed, 'If you think of anything else like this, come and tell me.' This was praise indeed and he later recalled it was on that day that he gained complete confidence in his ability to make a career in theoretical physics.

While in Göttingen, where he had hoped to work with Max Born – an opportunity that did not materialise owing to Born being in poor health – Mott received an offer of a lectureship at Manchester University under W. L. Bragg, which he accepted. A course of lectures he gave there on wave mechanics led to his first book, *An Outline of Wave Mechanics*, written at age 25. It was at Manchester where his interest turned to the properties of materials, stimulated no doubt by the power of the new technique of X-ray crystallography to determine the structure of solids, a method which earned W. L. Bragg (later Sir Lawrence and Cavendish Professor of Experimental Physics) the Nobel prize for physics, jointly with his father W. H. Bragg.

An invitation to return to Cambridge came in 1930 and Mott accepted a

Nevill age 18.

Nevill seated with Japanese scientists.

Fellowship at Gonville and Caius College. Now married, he found that theoretical physicists were rather more respected than when he had first entered the Cavendish. Nevertheless, the prime interest there was still in atomic physics and 1932 was the *annus mirabilis* of the Cavendish, when the neutron was discovered, the atom split and the positron (positive electron) was first observed, as predicted by Dirac. Stimulated by this extraordinary atmosphere and a suggestion by Fowler that he should write a book on atomic collisions, Mott collaborated with Harrie Massey to produce *The Theory of Atomic Collisions*, a classic tome which ran to several editions.

Bristol

In 1933 Mott accepted his first chair, Professor of Theoretical Physics at Bristol University. For six years up to the outbreak of the Second World War, he played a significant role in establishing Bristol as one of the foremost centres for solid-state physics. His philosophy, which he never abandoned during the remainder of his life, was to have theoreticians and experimentalists working closely together. Although many today see the advantages of such a liaison, it has never been easy to implement such a policy. Mott's ability to draw theory and experiment together, to persuade practitioners on both sides to work on common problems – indeed to blur any distinction between the two approaches – was one of his hallmarks for which he will be remembered. At Bristol he deliberately avoided setting up separate departments for theorists and experimentalists, a distinctive approach compared with other centres of learning.

Great success had been achieved in Europe, particularly in Germany, by the application of quantum mechanics to explain the properties of metals and the difference between metals and semiconductors. The work of Sommerfield, Bloch, Peierls and Bethe was especially successful and Mott set about continuing this tradition. He found an enthusiastic supporter in Harry Jones, a former research student of Fowler's, and together they wrote their famous textbook, *Theory of the Properties of Metals and Alloys*. Strange as it may now seem to the hundreds of solid-state physicists who learned their basic quantum mechanics of solids from this work, the book was not that well received by other theorists, being criticised for lack of rigour. What in fact we see in this book is mathematics applied to the direct interpretation of experimental data, interspersed with bold intuitive ideas which leapfrog the formal development when the rigorous mathematical treatment fails. This type of pragmatic approach – annoying to some pure theoreticians but beloved by experimentalists anxious to have formulae with which to analyse their data – was to be seen again during Mott's career.

From recent accounts of members of the Bristol research group, Mott provided a stimulating atmosphere, which they recall with fond memories. At miniconferences, which he organised with flair and a great sense of timing,

Nevill in the early 1980s.

participants were encouraged to perform at their best and many recall the excitement at these meetings. Numerous problems and puzzling properties of metals were solved at Bristol during those heady days, advancing the subject and providing much benefit to industrial metallurgists.

In those six prewar years at Bristol, Mott did not confine his attention to understanding the physics of metals. He began his first work on semiconductors, a field to which he was to remain attached for many more decades – in fact well into his retirement. The initial stimulus came from Ronald Gurney, who had arrived from Manchester and who, like Mott, had a remarkable talent for visualising solids in terms of their constituent atoms and electron waves, without the need for detailed mathematics. Together they wrote *Electronic Processes in Ionic Crystals,* another 'Mott and X' book (where in this case X was Gurney). Together they laid the foundation of the field of colour centres in alkali halides by providing a description of the defects involved in terms of negative ion vacancies. They also worked out the physics behind the photographic process – why light falling on a grain of silver bromide was able to produce a speck of silver – the latent image. In 1940 Mott was awarded the Harker and Driffield medal from the Royal Photographic Society for this work. Four years earlier, at the age of 31, he had been elected a Fellow of the Royal Society.

Mott's personal qualities and humanitarianism were evident after the German occupation of the Sudetenland, when there was a movement to rescue children of Jewish descent from Czechoslovakia. He and his wife, Ruth, with help from his sister, Joan, housed two young refugees, daughters of a Jewish musician. Lilly and Ilse Spielmann, who stayed in England after the war, recall their gratitude to this day.

The War and Return to Bristol

During the Second World War, Mott was involved in various defence projects relating to radar, shell fragmentation, the deployment of searchlights and to mathematical research on armaments generally. He succeeded Blackett as scientific adviser to a commanding officer in 1940. When he discovered that the number of enemy aircraft destroyed by British fire was being overestimated by quite a large factor, his commander told him that he must not disclose his findings until after the war was over. He wrote a paper on why German shells, which were made of steel with a high carbon content, fragmented into smaller pieces than the British shells. Apparently, memoranda on this still survive in American ordnance laboratories. At about the same time, the Royal Society awarded him their prestigious Hughes medal for his work before the war.

In the last days of the war, Mott was offered chairs in metallurgy and in theoretical physics at Cambridge, but an assurance that if he returned to Bristol he would succeed to Tyndall's position as head of department attracted him back there. He assumed the headship in 1948 and immediately set about appointing new staff. Many of the scientists at Bristol before the war, e.g. Skinner, Gurney, Harry Jones, Heitler and Fröhlich, took positions elsewhere and did not return. In their place Mott engaged, among others, Jack Mitchell (now occupying a chair at the University of Virginia), who demonstrated that silver in the photographic process precipitated preferentially along dislocations; Charles Frank, who produced novel theories of dislocations and crystal growth and predicted what became known as Frank–Read sources; Dirk Polder, who was a senior theorist; Jacques Friedel, who married Mott's sister-in-law, before returning to France to become leader of solid-state physics in Paris; and Nicolas Cabrera, who worked with Mott on a theory of why aluminium and stainless steel do not rust. All these workers by their individual

Nevill with his famous briefcase at a conference in Garmisch-Partenkirchen (1991).

and joint efforts once again made Bristol Physics Department into a leading research group. In 1949 Mott published his first paper on metal–insulator transitions, a topic that retained his interest to the end of his life. This was the period which saw the invention of the transistor at the Bell Telephone Laboratories in America and Mott took a keen interest in this development. He was later to propose a theory for what became known as Mott–Schottky barriers in semiconductors.

Administrative appointments slowed down Mott's personal research towards the end of his Bristol period (i.e. up to 1954). He became Dean of the Faculty of Science, served on government committees and was appointed

President of the Physical Society, overseeing its amalgamation with the Institute of Physics. He also took over the editorship of the *Philosophical Magazine* and became Chairman of the Board of Taylor & Francis, its publisher. Another activity which continued to occupy Mott's time for many years was his campaigning against UK development of the atomic bomb; he became chairman of the Atomic Scientists' Association, whose aims were to explain the true facts of nuclear energy to a wide audience and to investigate proposals for its control.

Cavendish Professorship

In 1953 Mott received an offer of the Cavendish Chair of Experimental Physics at Cambridge, following Sir Lawrence Bragg's retirement one year earlier. He recalls in his autobiography that he was not too happy at the thought of leaving Bristol but felt he could not turn down the offer to follow in the footsteps of James Clerk Maxwell, Lord Rayleigh, J. J. Thomson, Ernest Rutherford and Lawrence Bragg, the previous holders of this prestigious chair. It might seem strange that the title of Professor of Experimental Physics should be offered to a theorist but, apart from Rutherford, none of the previous incumbents had been principally experimentalists. One of Mott's first decisions, taken even before he actually took up the post, was to halt plans for the construction of a linear accelerator for high-energy particle physics – not because he was against this type of research but simply because he did not believe the machine would be able to compete against similar American installations.

The move back to Cambridge in 1954 was Mott's last; he was to remain working there not only up to his official retirement in 1971 but for another twenty-five years thereafter. On his appointment, Professor Fred Seitz, the American solid-state physicist, wrote:

Dear Mott

For several weeks there has been a consistent report that you have accepted the appointment as director of the Cavendish Laboratory. Deepest congratulations. During our sojourn in Tokyo I made bets with our colleagues that this would occur, since it was the most logical solution of the issue. Either my judgement is lucky or very profound.

Undoubtedly , this means that you will be very busy during the summer and that we will catch only fleeting glimpses of you at the various meetings. In any case, we all hope that Bristol will manage to hold its level and remain the very fine center of research which you have made it.

Sincere regards
Frederick Seitz

That Cambridge was a world apart from Bristol became clear on Mott's introduction at the Cavendish. His first contact was with the Administrative Secretary of the department. 'What are your problems?' Mott asked of him. 'Well,
Prof., I did want to talk to you about the Cavendish cricket match, academic
staff versus assistants,' came the reply!

Administrative duties and matters unrelated to his own research occupied
Mott's attention for a great deal of his time as Cavendish Professor. Reform of
the Natural Sciences Tripos, which permitted undergraduates to specialise in
physics at an earlier stage of their studies than had been possible previously,
was one of his missions. Another was to change the physical sciences curriculum in schools through the Nuffield Foundation. In chairing an advisory
committee of that body, he had to steer a difficult course between the prota-
gonists of 'learning by doing', which involved students being encouraged to
discover the laws of physics for themselves, mainly by experimental physics,
and ideas based on his own education, in which the laws of physics are
approached through mathematics, their validity and beauty becoming clearer
when expressed and taught this way. A compromise was called for, and to a
large extent a successful solution was found. As a member of the General
Board and the Council of Senate, both influential bodies of the University,
Mott steered through important decisions, not least the creation of the first
science park by Trinity College, which attracted science-based industries to the
city. Most UK universities now have their own science park.

In 1959 Mott was elected Master of Gonville and Caius College. His
predecessor, Sir James Chadwick, was a distinguished nuclear physicist who
had discovered the neutron in 1932 and who had been largely responsible for
Anglo-American cooperation in the Manhattan Project during the Second
World War. Chadwick had retired prematurely following dissent within the
College Fellowship and Mott was persuaded that he could heal the divisions
that had developed. His appointment, as with Masterships of other Cambridge
colleges, did not require him to resign the Cavendish Chair, but the task of
holding the two positions must have been arduous. Life in Caius tended to
centre around the Master's Lodge, a splendid place to entertain the many
distinguished visitors. The lodge is situated within the college precincts,
opposite the chapel. Parts of it date from the sixteenth century and, in addition to wonderful dining and drawing rooms, it contained, when Mott lived
there, five bathrooms and a hatch leading into the college kitchens.

C. P. Snow's novels describe, with much reality, the intrigues and politics
of college life, particularly in relation to the appointment of the Master. Mott
used to say, 'He didn't know the half of it!' After six years of service, he resigned from the Mastership, having become tired of the petty wranglings of some
of the Fellows, just as his predecessor and indeed the founder of the college,
John Caius, had done also. But this was not until after he had reformed the
college's admissions policy (at Cambridge, students are admitted to the university via the colleges), and arranged for the first time that Fellows could
bring in lady guests to dine in hall. During his tenureship, Mott continued his

interest in control of nuclear weapons and hosted a Pugwash conference in the college, at which he recalled Henry Kissinger arguing with the Russian representatives on their hostility to China. It was also during his Mastership of Caius that he was awarded his knighthood.

Non-Crystalline Solids

When the present writer arrived in Cambridge in 1967, Mott was still Head of the Cavendish and had once again begun to turn his attention seriously to solid-state physics. His interest in metal–insulator transitions had never waned. When I described to him work I had been doing at the University of Illinois on heavily doped semiconductors, which behave as metals when the concentration of donors or acceptors exceeds a critical value, he displayed an interest far beyond my expectations, particularly as I had not been the first to make such measurements; Professor Fritzsche in Chicago and others had undertaken similar and more extensive studies. What Mott saw, with his customary and extraordinary insight, was that the metal–insulator transition in doped semiconductors was intimately related to a theoretical paper by Phil Anderson on the effect of disorder on the electronic states in a solid.

Anderson's classic paper of 1958 (which he later described as 'one that is often quoted but seldom read') showed how disorder 'localises' electronic states, distinguishing them from band-like 'extended' states. A metal–insulator transition occurs when the Fermi level passes through the energy separating extended from localised states. The disorder, in the case of heavily doped semiconductors, arises from the random positions of the donors and from the electric fields associated with charged compensating acceptors. Mott was able to show that the critical concentration of dopants for the 'Anderson transition' was not dissimilar to that for the 'Mott transition', the theory of which was based on a screening argument. In considering electrical conduction on the insulating side of the transition, he formulated what is now known as the Mott $T^{-1/4}$ law, which describes how carriers hop between localised states over a distance that is temperature dependent, so-called variable-range hopping. I recall his making a back-of-the-envelope derivation of this relationship, a calculation which was subsequently done 'properly' but with essentially the same result. But these were only the beginnings of a decade or more of activity on other disordered systems – especially amorphous or non-crystalline semiconductors – which eventually led to Mott's 1977 Nobel prize for physics, shared with Anderson and Van Vleck.

It was an illuminating experience to see how Mott developed and nurtured the field of amorphous semiconductors. At the time, few groups showed any interest in them. Although the Xerox Corporation in America had achieved great success with their first 'dry' photocopying machine, which used amorphous selenium as the photoreceptor, the physics behind the process was not understood. In the Soviet Union, Kolomiets had been working on the properties of glasses; in Germany, Stuke had a small group and, in the United Kingdom, Spear was studying transport in thin films of germanium and silicon. What Mott did was to bring together these disparate activities using his by now well-tested methods. Pouncing eagerly on new results, he formulated his ideas and communicated them via his familiar handwritten letters to interested parties. He organised miniconferences (in the way he had done at Bristol), visited laboratories for personal discussions, suggested PhD topics, and wrote draft papers for wide circulation and comment. Thereby he rapidly became the father figure of a growing community and the Cavendish became the conduit through which ideas were channelled.

During a visit to America he was introduced to Stan Ovshinsky, who had started a small company, Energy Conversion Devices (ECD), in Detroit, to exploit the potential of non-crystalline materials. Ovshinsky had developed two forms of electronic switch based on chalcogenide glasses, largely dismissed by other companies as being of little importance. But Mott thought otherwise and attempted to explain their behaviour in terms of his new concepts on conduction in disordered materials. The possibility of commercial utilisation of amorphous semiconductors (whether it was to be fulfilled or not was of little consequence) had a dramatic influence on the growth of the field.

The third conference in the series International Conferences on Amorphous and Liquid Semiconductors (held every other year since 1965) took place in Cambridge and the number of participants was four to five times that of the previous meeting. Mott chaired most of the sessions, giving his own views and interpretation of virtually every paper presented. The concepts of mobility edges, hopping conduction, minimum metallic conductivity, the '8-N rule', relaxation of the k-selection rule, tail states, dangling bonds, etc. were added to the vocabulary of practitioners in the field; it would not be too generous to give Mott credit for introducing most of these ideas. Another significant event in the development of the subject was the discovery by Spear's group of the way to dope amorphous silicon, providing a further boost to worldwide interest and leading to the development of solar cells, thin-film transistors and other devices now used extensively in the electronics industry. Last year (1997) saw the seventeenth conference in the same series mentioned above, the only change being to its name – it is now the 'International Conference on Amorphous and Microcrystalline Semiconductors', but it still started with the 'Mott lecture'.

Mott's work on amorphous semiconductors was not interrupted by his retirement in 1971. The same year saw the first edition of *Electronic Processes in Non-Crystalline Materials*, co-authored with the present author. A second edition, extensively rewritten, appeared in 1979. In the intervening years he had also written *Metal–Insulator Transitions* (1974) and, for sixth-formers, *Elementary Quantum Mechanics* (1972). When the Cavendish Laboratory moved from its old home in Free School Lane to its present site in West Cambridge, Mott collaborated with Abe Yoffe and Mike Pepper in their respective studies of low-dimensional crystals and silicon inversion layers, as well as with numerous long-stay visitors who were attracted to Cambridge by his late bursts of scientific activity and the interest he showed in their work.

Final Decades

During the late 1980s Mott turned his attention to another major field of scientific discovery, high-temperature superconductivity. In this endeavour he collaborated principally with Sasha Alexandrov, a Russian theoretician for whom Mott was able to find a position in the Cavendish (financed first by the Leverhulme Trust and subsequently by the Newton Trust and Gonville and Caius College), and with whom he wrote two more books. The field is still too new to assess Mott's contributions to the subject, but no matter how they are judged in the future, he continued to stimulate and inspire colleagues, both young and old. His unforgettable manner and his wise counsel continued until well past his ninetieth birthday – an astonishing achievement.

Throughout his long career, Mott's energy and enthusiasm pervaded much of twentieth-century physics. His appetite for new challenges seemed inexhaustible and his contributions to science will be quoted well into the next century. But what cannot be so readily recorded is the influence he had on the lives and careers of numerous friends and colleagues, people he advised and guided with such great generosity. It is hoped that this book will go some way towards ensuring that the personal reminiscences and appreciations of many of his closest acquaintances are recorded for posterity.

Edward A. Davis
Professor of Experimental Physics
University of Leicester

List of Contributors

Professor J. V. Acrivos	San José State University, USA
Professor A. S. Alexandrov	Loughborough University
Professor J. F. Allen	University of St Andrews
Professor P. W. Anderson	Princeton University, USA
Mr D. W. B. Baron	Formerly of the National Extension College, Cambridge
Professor K-F. Berggren	Linköping University, Sweden
Mr M. Berry	Colleague in education
Professor R. J. Blin-Stoyle	University of Sussex and colleague in education
Professor L. M. Brown	University of Cambridge
Sir Clifford Butler	Formerly of Loughborough University
Professor R. W. Cahn	University of Cambridge
Professor Y. Cauchois	Université Pierre et Marie Curie, Paris, France
Professor B. R. Coles (deceased)	Imperial College, London
Professor Sir Alan Cottrell	University of Cambridge
Mrs A. Crampin	Nevill's daughter
Professor C. Crussard	Formerly of Pechiney, Neuilly, France
Professor R. H. Dalitz	University of Oxford
Mr T. Dalyell	Member of Parliament
Professor R. J. Eden	University of Cambridge
Professor P. P. Edwards	University of Birmingham
Professor Sir Sam Edwards	University of Cambridge
Dr S. Elworthy	Oxford Research Group
Professor J. E. Enderby	Formerly of University of Bristol
Mrs E. Ferguson	Taylor & Francis
Miss S. Fieldhouse	Nevill's former secretary
Mrs P. Fisk	Family friend
Mrs J. Fitch	Nevill's sister
Sir Charles Frank	University of Bristol
Professor J. Friedel and family	Université de Paris-Sud, France
Dr and Mrs L. Friedman	US Air Force, Rome Laboratory, Mass., USA
Professor and Mrs H. Fritzsche	Formerly of University of Chicago, USA
Mr D. F. Gibbs	Formerly of University of Bristol
Mrs L. Gill	Family friend

Professor J. B. Goodenough	University of Texas at Austin, USA
Professor G. N. Greaves	University of Wales, Aberystwyth
Professor Dr F. Hensel	Universität Marburg, Germany
Sir Peter Hirsch	University of Oxford
Dr J. P. Horder	Nevill's brother-in-law
Professor A. Howie	University of Cambridge
Lady Jeffreys	Girton College, Cambridge
Professor H. Kamimura	Science University of Tokyo, Japan
Professor M. Kaveh	Bar-Ilan University, Israel
Professor D. Kuhlmann-Wilsdorf	University of Virginia, USA
Miss E. C. Leschke	Family friend
Professor W. Y. Liang	IRC in Superconductivity, Cambridge
Professor Sir Bernard Lovell	University of Manchester
Sir James Menter	Formerly of Queen Mary College, London
Sir John Meurig Thomas	University of Cambridge
Professor J. W. Mitchell	University of Virginia, USA
Sir William Mitchell	Wadham College, Oxford
Rt Revd H. Montefiore	Former Bishop of Birmingham
Professor K. Morigaki	Hiroshima Institute of Technology, Japan
Professor E. Mytilineou	University of Patras, Greece
Professor F. R. N. Nabarro	University of the Witwatersrand, South Africa
Dr A. D. I. Nicol	University of Cambridge
Mr S. R. Ovshinsky	Energy Conversion Devices, Troy, USA
Professor M. Pepper	University of Cambridge
Dr M. F. Perutz	MRC, Cambridge
Professor Sir Brian Pippard	University of Cambridge
Revd John Polkinghorne	Formerly of Queens' College, Cambridge
Professor M. Pollak	University of California Riverside, USA
Mrs C. Pongratz-Lippitt	Family friend
Professor T. V. Ramakrishnan	Indian Institute of Science, Bangalore, India
Professor C. N. R. Rao	Jawaharlal Nehru Centre, Bangalore, India
Professor T. M. Rice	ETH Zurich, Switzerland
Professor N. Rivier	Université Louis Pasteur, Strasbourg, France
Professor J. Rotblat	University of London, UK
Professor C. Schlenker	CNRS and Institut National Polytechnique de Grenoble, France
Mrs E. Scott	Family friend
Professor Dr A. Seeger	MPI für Metallforschung, Germany
Professor F. Seitz	The Rockefeller University, USA
Professor D. Shoenberg	University of Cambridge
Professor I. N. Sneddon	University of Glasgow
Professor W. E. Spear	Formerly of University of Dundee
Mr E. G. Steen	Relation

Dr R. A. Street	Xerox Palo Alto Research Center, USA
Professor J. Stuke	Formerly of Universität Marburg, Germany
Professor D. Tabor	University of Cambridge
Dr M. L. Thèye	Université Pierre et Marie Curie, Paris, France
Dr G. A. Thomas	Bell Laboratories, Lucent Technologies, USA
Sir Arthur Vick	Formerly of Queen's University, Belfast
Dr P. K. Walker	Former Bishop of Ely
Dr A. D. Yoffe	University of Cambridge
Lord Young of Dartington	Formerly of the Open University
Professor J. Ziman	Formerly of University of Bristol

Chronology of Nevill Mott's Life and Work

1905	Born 30 September
1923	Major scholarship to St John's College, Cambridge
1929–1930	Lecturer at Manchester University
1930–1933	Fellow and lecturer, Gonville and Caius College, Cambridge
1930	Married Ruth Horder
1930	Publication of *An Outline of Wave Mechanics*
1933	Publication of *The Theory of Atomic Collisions* (with Massey)
1933–1948	Melville Wills Professor of Theoretical Physics, University of Bristol
1936	Fellowship of the Royal Society
1936	Publication of *The Theory of the Properties of Metals and Alloys* (with Jones)
1940	Publication of *Electronic Processes in Ionic Crystals* (with Gurney)
1941	Hughes Medal of the Royal Society
1948	Publication of *Wave Mechanics and its Applications* (with Sneddon)
1948–1954	Henry Overton Wills Professor of Physics, University of Bristol
1951–1957	President of the International Union of Physics
1952	Publication of *Elements of Wave Mechanics*
1953	Royal Medal of the Royal Society
1954	Corresponding member, American Academy of Arts and Sciences
1954–1971	Cavendish Professor of Experimental Physics, University of Cambridge
1955	President of Modern Languages Association
1956	Publication of *Atomic Structure and the Strength of Metals*
1956–1958	President of the Physical Society
1957	Foreign Associate of the United States National Academy of Sciences
1957–1960	Governing Board of the National Institute for Research in Nuclear Science
1959	Crowther Committee on Education from 16 to 18
1959–1962	Chairman of the Ministry of Education's Standing Committee on Supply of Teachers
1959–1962	Academic Planning Committee and Council of University College of Sussex
1959–1966	Master of Gonville and Caius College, Cambridge
1961–1973	Chairman of Nuffield Foundation's Committee on Physics Education

1962	Hosted first UK Pugwash conference
1962	Knight Bachelor
1964	Honorary Member of Akademie der Naturforschung Leopoldina
1965–1971	Physics Education Committee (Royal Society and Institute of Physics)
1970	Honorary Member of Société Française de Physique
1970	Grande Médaille de la Société Française de Métallurgie
1970–1975	Chairman of the Board at Taylor & Francis
1971	Honorary Fellow at St John's College, Cambridge
1971	Publication of *Electronic Processes in Non-Crystalline Materials* (with Davis)
1971–1973	Senior Research Fellow, Imperial College, London
1972	Copley Medal of the Royal Society
1972	Publication of *Elementary Quantum Mechanics*
1972	Honorary Fellow of the Institute of Physics
1973	Faraday Medal of the Institute of Electrical and Electronics Engineers
1974	Publication of *Metal–Insulator Transitions*
1975	Chairman of the Science Park Committee, University of Cambridge
1975	Honorary Member of the Institute of Metals, Japan
1975	Honorary Fellow of UMIST
1976–1986	President of Taylor & Francis
1977	Fellow of Darwin College, Cambridge
1977	Chevalier, Ordre Naturel du Mérite, France
1977	Nobel prize for physics (jointly with Anderson and Van Vleck)
1979	Second edition of *Electronic Processes in Non-Crystalline Materials* (with Davis)
1980	Honorary Member of Sociedad Real Española de Fisica y Quimica
1980–1983	Adrian Visiting Fellow, University of Leicester
1982	Foreign Fellow, Indian National Science Academy
1985	Honorary Member of the European Physical Society
1986	Publication of *A Life in Science* (his autobiography)
1986	Publication of *Conduction in Non-Crystalline Materials*
1990	Second edition of *Metal–Insulator Transitions*
1991	Publication of *Can Scientists Believe?*
1993	Second edition of *Conduction in Non-Crystalline Materials*
1994	Publication of *High Temperature Superconductors and Other Superfluids* (with Alexandrov)
1995	Publication of *Sir Nevill Mott: 65 Years in Physics* (with Alexandrov)
1995	Companion of Honour
1996	Publication of *Polarons and Bipolarons* (with Alexandrov)
1996	Died 8 August

Honorary Doctorates

Louvain	London	William and Mary
Grenoble	Warwick	Stuttgart
Paris	Lancaster	Marburg
Poitiers	Heriot-Watt	Bar Ilan
Bristol	Oxford	Lille
Ottawa	East Anglia	Rome
Liverpool	Bordeaux	Lisbon
Reading	St Andrews	Cambridge
Sheffield	Essex	Linköping

Prewar

A Happy Childhood

Joan Fitch

Nevill was born in a nursing home at Leeds, but the home to which his mother must soon have brought him was Tems Cottage, beside the Tems Beck in the village of Giggleswick in the West Riding of Yorkshire. Behind the cottage, across a field, was the workhouse, and on the hill above stood 'Daddy's Chapel', the curious domed Gothic building given to Giggleswick School by that Mr Morrison whose fortune was made by a corner in crêpe just in time for the death of the Prince Consort.

Daddy was Francis Mott, First Class Honours in Natural Science at Cambridge, and at the time of Nevill's birth senior science master at Giggleswick School. Mummy's academic record was equally distinguished; she had been equal to Eleventh Wrangler when there was still an order of merit in the Mathematics Tripos and women were not allowed in the regular list. Francis and Lilian Mary had met in the Cavendish Laboratory, both engaged in physics research. There are photographs in the laboratory's museum in which they both appear, she wearing a ridiculous hat. Little did they know that their first-born would one day be Cavendish Professor of Physics!

Francis's salary was something in the neighbourhood of £300 a year. Out of that they were able to employ a full-time live-in nanny and a full-time, though non-resident maidservant; such were prices in the early 1900s. Middle-class wives in those days did not work, or not for money; but Mary was able to help the school in various ways and she also did some voluntary social work, as she did throughout her married life. She was one of those for whom 'the miseries of the world/Are misery, and will not let them rest.' She was also a great advocate of votes for women; and though never a militant suffragette, was once visited by the great Mrs Pankhurst.

I was born when Nevill was nearly 2, and there is no record that he was ever jealous of his little sister. He was sorry for me when I was sore from vaccination, and when I became old enough to play, we became good playmates. Nana, Alice, was much loved by us. It was with her that we went for walks, up the hill to Daddy's Chapel and beyond, to Settle across the river, or best of all, to the ebbing and flowing well. The well still ebbs and flows, but no one now sees what we saw, the Silver Cord, a long bubble of air stretching from one side of the well to the other, occurring at some stage of the syphon system which causes the ebb and flow. Here was Nevill's first scientific problem, Why does it do that? Other early questions are, Daddy, when I look in that looking-glass, why do I look big? Daddy, how could God make the

Nevill Francis Mott with his mother (age 3 months).

world if it hasn't an end? If water is a liquid, does that make treacle a slow liquid?

Giggleswick lies amid glorious limestone scenery, which has delighted us in later life. But it is too vast and too austere for small children. Our next home was ideal. Brocton is a village a few miles out of Stafford, where Francis was now Assistant Director of Education, and The Poplars is almost the last house in the village on the lane up to Cannock Chase. We could walk out on to the Chase in minutes. Cannock Chase is a wilderness of heather, gorse and bracken; there is an oak wood and a fir wood, and a small intimate valley with a brook where we could make dams. Just above the house was a sandpit where we were allowed to go and play for hours together. Was there no danger in those days when letting children out alone? The modern child has no such liberty.

Nevill gives sister Joan a wheelbarrow ride (age 8).

Nevill age 11.

Of course life was not all play; Nevill was now 6 and there were lessons to be learnt. Our mother taught us for an hour every day. Nevill must have been miles ahead of me, not only because he was older, and cleverer, but also because he was good and I was bad. Many times I stumped out of the room in a temper; but Nevill would say, 'Leave Joanie to me,' and his reasonableness would calm me down.

Nevill began to invent machines. We had books of drawing-paper in which I drew picture stories, but he drew plans for machines. Though none was ever constructed, and probably wouldn't have worked if it had been, it showed a dawning interest in technology and the beginnings of science.

Our idyllic existence at Brocton was brought to an end, first by the outbreak of war in 1914, then by the need to go to school. The War Office built a camp on Cannock Chase, and our quiet lane became a roaring highway. We moved into Stafford, and when Nevill was nearly 10 we began our schooling. He was sent to a newly opened prep school, Baswich House, first as a weekly boarder but later as a full boarder, because the boys teased him about running home. Being no good at games, and what the boys no doubt called a swot, he was always something of a misfit at school, both at Baswich House and at the public school, Clifton, to which he went later on. Why our parents, so modern in many ways, condemned him to this long purgatory, we never could make out; but no doubt they wanted to do the best for their son. Indeed, Clifton was noted for science, and the mathematics master was a first-rate teacher. Nevill never complained, but bore it stoically to the end, when he won a scholarship to St John's College, Cambridge, and the bliss of freedom.

MRS JOAN FITCH, Nevill's sister, was born in 1907 and has lived in Cambridge since 1957. She graduated from Newnham College in mathematics and economics in 1928. She has kept in close contact with the family, and Nevill used to stay with her when he travelled from Aspley Guise to work at the Cavendish Laboratory after his official retirement.

Cambridge and Manchester 1924-1933

Bertha Swirles Jeffreys

In 1924, when Nevill Mott came up to St John's to read mathematics, I had taken Part II of the Tripos and was embarking on Part II physics. He once said to me, 'Of course one had done most of the Part I stuff at school.' I had not, though I was well taught. He compressed a three-year course into two years and in his second year was attending advanced lectures. I remember him at Ebenezer Cunningham's in St John's on electron theory and at some stage he and I were the entire class for T. M. (later Sir Thomas) Cherry in Trinity on the solution of dynamical equations. This made it difficult to cut.

In 1926 Nevill was a Wrangler with Distinction in Schedule B. The Mayhew prize for the best performance in applied mathematics was shared between J. A. Gaunt (1904–1940) and A. H. (later Sir Alan) Wilson, both third-year men. I wrote about Gaunt in *Notes Rec. R. Soc. Lond.* (1990) and there will be a Royal Society Memoir of Sir Alan.

In January 1926 the Girton Mathematical Society invited Mr Cunningham to give a lecture to them along with members of the St John's Adams Society. His title was 'Mathematics and Morals', published in the *Mathematical Gazette*, Volume 13 (1927). A year later I was the speaker at a joint meeting in St John's. My subject was the prelude to wave mechanics. During that summer I happened to meet Nevill in the Cavendish, and as he was the president-elect, he asked me if Girton was going to invite them again. I did not know as I was going to Göttingen for the winter semester. He gave me an excellent lunch, ordered on the spot from the college kitchen, in his rooms in Third Court. Used as I was to the simpler ways of Girton, I was greatly impressed. Nowadays in St John's it would be self-service in the buttery dining room.

When I returned from Göttingen to Cambridge in the following April, Nevill wanted to know all about it as he was keen to go abroad. He had, I think, already proved the Rutherford scattering formula according to wave mechanics. It is worthwhile to record that Sir Charles Darwin, Rutherford's mathematical assistant in Manchester, told me that Rutherford himself produced the elegant proof according to classical mechanics.

In the autumn of 1928 Nevill and the Hartrees went to Bohr's institute in Copenhagen and I became an assistant lecturer in the Mathematics Department of the University of Manchester. Nevill met George Gamow and fairly soon I had an excited message about this Russian with a theory of α-disintegration. It was years later that I found how slapdash his mathematics was, but in principle the theory worked.

In the Long Vacation of 1929 I spent some time in Cambridge. In succession to E. A. Milne, Douglas Hartree had been appointed Beyer Professor of Applied Mathematics in Manchester and W. L. Bragg had invited Nevill to a special lectureship in the Physics Department. Nevill was engaged to Ruth Horder, who had to complete another year on Part II of the Classical Tripos at Newnham College. Jocelyn Toynbee, her director of studies, admonished me to 'keep that young man in Manchester!' In this I was partially successful. At the end of the Lent term, Nevill asked R. W. James and me to go to a register office in Cheetham Hill as witnesses of their marriage. After a brief ceremony we all went to lunch at a Squirrel restaurant. Ruth returned to Cambridge for the Easter term and obtained first-class honours in Part II of the Classical Tripos, as expected.

Nevill lived at the Manchester University Settlement in Ancoats and took the part of an old man in their Christmas play. Our transport was shanks's pony, tram or bus. We did not have cars or even jalopies. I was fortunate later to have lifts in the 'Hartree taxi' with the children.

At a university reception early in the academic year, Nevill was in full evening dress, a present from his parents. I introduced him to the philosopher Professor Samuel Alexander, who questioned us about absolute and relative rotation and Newton's bucket experiment. I rather scored as I had read about it in C. D. Broad's *Scientific Thought*. A day or so later Professor Alexander said to me, 'The most distinguished person I've met lately is your friend Mott.'

Nevill's lectures were given in the Physics Department at about five o'clock. Douglas Hartree, Sydney Goldstein and I went over from the Mathematics Department, up the stairs past the plaque to Moseley. Bragg, James, Nuttall, W. H. Taylor, Bell, Brentano, Scott Dickson and E. J. Williams are the physicists I remember. The lectures were published as *An Outline of Wave Mechanics* by the Cambridge University Press in 1930.

In the autumn of 1930 Nevill returned to Cambridge to a university lectureship and a fellowship at Gonville and Caius College. They settled in a house in Sedley Taylor Road. Gamow may have thought this rather bourgeois and he addressed a letter, 'Rooth and Nevill Mott, their own house . . . .' I visited them early in 1931 and Nevill took me to the Kapitza Club. The speaker was G. Beck on scattering of particles.

In part of my PhD thesis in 1928 I attempted a theory of the internal conversion of γ-rays. This was non-relativistic and did not agree with C. D. Ellis's experiments. In 1930 I met Casimir in Copenhagen. He had a go at the problem and, delighting in American slang, sent me a postcard, 'I've gotten quite a cute formula, but the result ain't large enough.' At the suggestion of R. H. Fowler, J. R. Hulme carried out a more detailed relativistic calculation and obtained agreement for some radium C lines. As I had done, he assumed that the outer electrons were subjected to the field of an electric dipole. With H. M. Taylor, Nevill worked on the theory and in particular they considered the effect of the field of a quadrupole. In 1931–32 I was at Bristol, and some of us went from there to the memorable Royal Society meeting, where we heard about the experiment of Cockcroft and Walton, and Chadwick's discovery of the neutron. I remember that Harold Taylor asked me what I knew about a quadrupole, which at that time was not much. Their papers were published in 1932 and 1933.

In 1933 Nevill became Professor of Theoretical Physics at Bristol and I returned happily to Manchester. At the Whitsuntide holiday in 1934 I visited a Girton friend in Bristol. The Motts invited me to lunch and Nevill was enthusiastic about the theory of metals. I think this was brought about by Harry Jones, who had been interested in the subject in 1931–32. For some years I stuck to atomic spectra.

During the war we lived in the Thatched Cottage near St John's. One evening Nevill paid us a surprise visit, and to his amusement, he found that Harold had removed his shoes and socks and was warming his feet by the fire. Actually, for wartime conditions, the cottage was quite warm.

When Nevill became Cavendish Professor in 1954, I was much involved in college teaching and college committees, and Harold was not far off retirement; this enabled us to travel. I felt really old when someone told me a story about Nevill, one that I first heard about J. J. Thomson: 'Which way was I going when I met you? Oh, then I have had lunch.'

BERTHE SWIRLES (Lady Jeffreys) was born in Northampton in 1903. She obtained her MA and PhD and was a scholar at Girton College, Cambridge. In 1938 she returned there as fellow and lecturer, after a lectureship at the University of Manchester. In 1940 she married Harold Jeffreys, and was joint author with him of Methods of Mathematical Physics. *Her research interests have been in theoretical atomic physics and related fields. She is a Life Fellow of Girton College.*

Memories of Nevill and his Bristol Team

Elisabeth Charlotte Leschke

There cannot be many people apart from his family who have known Nevill as long as I have, because we all came to Bristol in September 1933 – Nevill just before his twenty-eighth birthday, as Melville Professor of Theoretical Physics at Bristol University, and I just after my nineteenth, as a student teacher of German at Badminton School, where Ruth became a classics teacher. They befriended me, a young refugee from Germany, and over the years we became friends for life.

Badminton School also provided me with another link with the Physics Department because one of my first pupils was Elizabeth Tyndall, the younger daughter of Professor A. M. Tyndall, who was the Director of the Physics Laboratories and also the Chairman of the Board of Governors of Badminton. (He later became Acting Vice-Chancellor and it was he who, at the last Senate meeting he chaired, proposed the foundation of a Chair of Music.)

The Mott family lived in Stuart House – which later became the Vice-Chancellor's residence – opposite the Royal Fort, which housed the Physics Department we just called the Lab, and next to the Royal Fort House, where the Music Department was situated before it moved to the Victoria Rooms in 1996.

I feel sure Nevill would have liked me to mention some of the people who worked with him. Among the German refugees who came to the Lab in the thirties were three pairs of brothers: Heinz and Fritz London, Herbert and Ali Fröhlich, Walter and Hans Heitler. Both of the Fröhlich and Heitler brothers became professors, Herbert in Liverpool and Ali in Cambridge. Being the younger, though just as tall as his brother, Ali was known as Baby Fröhlich. And when they were both here, Hans Heitler was known as Walter's brother, but when Walter left Bristol for Dublin and Zürich and Hans stayed here as Professor of Mathematics, Walter became Hans' brother.

After the war, Nevill brought Charles Frank to Bristol in 1946. I met him and his wife, Maita, at Herbert Fröhlich's flat in Clifton. This makes me realise that we have been friends for half a century.

Among other mutual friends there was Dick Polder from Philips, Eindhoven, who like many of us enjoyed country walks. A very good cellist, he played chamber music with David Gibbs and others. He later became a professor in Utrecht.

Then there was Jack Mitchell, from New Zealand, for whom I checked German translations of his papers, including the one on photosensitivity after which he was awarded a Fellowship of the Royal Society. He later became a professor in Charlottesville, Virginia, where he joined Nicolas Cabrera, who had been in Bristol and Paris before he went to the States. I often visited Nicolas and Carmen in their flat in the sixth *arrondissement* of Paris. I saw them again, along with Jack, in 1962 when I was visiting internationally minded schools in America. But I went on visiting their flat many more times after they had left Paris because it was taken over by the Friedels.

When Jacques Friedel came to Bristol, he stayed with Ruth Mott's sister, Mary Horder, and her friend Diana; he married Mary in 1952. I was invited to their wedding at which the late Bishop Mervyn Stockwood officiated; the reception was held in the Royal Fort Gardens.

Some years later I was invited to another occasion in the extended Mott family, the wedding of the Friedels' younger son, Paul. It was a very impressive ecumenical marriage service conducted by a Protestant and a Catholic clergyman. Equally impressive, though in a wordlier way, was the reception held at the famous Haute Ecole Polytechnique, where I met Jack Mitchell again as he is Paul's godfather.

To return from the States and the Continent, back in Bristol I remember wonderful Christmas parties at the Lab when the staff let their hair down and performed highly entertaining pantomimes they had written. One of them started with Charles Frank and David Gibbs discussing what to perform that

Nevill outside the Bristol Physics Department (circa 1937).

11

year. Charles: 'I only know one pantomime and that is Cinderella.' David: 'We've done that one.' On another evening, Professor Tyndall appeared as Father Christmas sitting at his desk answering the telephone. 'Buckingham Palace? (*Aside*) I wonder if they are going to make her the Duchess of Henleaze?' 'Oh, hello George!'

I don't quite remember which complicated experiment Nevill was demonstrating another time, but I have a vague idea that it was something like a thermometer going down when it was heated – I wonder if anyone remembers this? Talking about thermometers reminds me of Professor Tyndall telling us that, as a nervous young lecturer, he was explaining thermometers to his students; having dealt with Fahrenheit and Celsius, he mentioned an obscure scale called Réaumur which went from eight to naughty, and he wondered why the students laughed.

The link between Professor Tyndall and Nevill survived Tyndall's death. I have a touching example of this because Nevill often addressed his letters to me in Henleaze Gardens, where the Tyndalls used to live, instead of Henleaze Road, where I live.

When the Motts went to Cambridge, I often stayed with them, first in Sedley Taylor Road and later in the Master's Lodge at Gonville and Caius College. On one of these visits I brought them a little turntable for their breakfast table and Nevill promptly put their cat on it. He had a lovely sense of humour and great charm.

In June 1960 there was a conference at Caius, and because it was attended by a lady physicist, other females could be invited to the dinner. I remember meeting Professor Bob Chambers from Bristol on the train to Cambridge. The dinner was a very formal one, served by a white-gloved butler. The guests of honour were the President of the College, Dr Joseph Needham, his wife Dorothy and the Dean of the College, the Rt Revd Hugh Montefiore. Earlier in the day, Nevill said to Ruth: 'Shall we tell Liselotte?' and proceeded to enlighten me about some of the stories behind the guest list. It was pure C. P. Snow! In fact, as I read in Anthony Tucker's obituary in the *Independent*, Nevill felt just like that, too. He writes that Nevill retired as Master of Caius because of his contempt for the trivial politics of college life. 'These,' he wrote in his autobiography, *A Life in Science* (1986), 'are far worse than anything described by C. P. Snow.'

Nevill was not just an outstanding scholar but a truly remarkable man in every way. If I may paraphrase Orwell: All men are special but some men are more special than others. Vastly superior though he was, Nevill never made me feel inferior but always treated me as an equal and a friend, for which I shall be eternally grateful. The very fact that we were using Christian names in the early thirties makes me marvel in retrospect.

I was reminded of Nevill when I recently read this: 'There are a few rare people in this world whose lives take flight, who make their moment on this earth count as they soar above the rest of humanity.' Nevill Mott was one of them.

MISS E. C. LESCHKE (Liselotte to her friends) was born in Berlin and educated in Hamburg. She came to Badminton School, Bristol, as a student teacher in 1933 and in 1935 she began reading modern languages at Bristol University. She graduated in 1938 and in 1942 she returned to Badminton School as a full-time teacher and librarian until her retirement in 1974. Living in Bristol enabled her to work on various University Committees, including 18 years of editing the Alumni Gazette *and representing Bristol on the Committee of the Conference of University Convocations. She is still on the Bristol Convocation Committee and a member of Court. In 1996 she was awarded an Honorary M.A. for her services to the University.*

Bristol: First Years

Bernard Lovell

In 1933, when Nevill Mott came to the University of Bristol as the Melville Wills Professor of Theoretical Physics, there were six of us in the final honours year of the physics course, and Mott's impact on that small group was dramatic. So far our education in theoretical physics had been in the hands of Lennard-Jones,* who had filled the blackboard with the seemingly endless equations of classical physics. We had gone deeply into Maxwell's equations then one day, to our surprise, a young man walked into the lecture room. He swept the chalk of those equations from the board and with scarcely a word wrote

$$\lambda = 2l/n$$

which he said was the wavelength of the de Broglie waves for an electron 'shut up in a box', l being the distance between the walls and n an integer. That was our introduction to wave mechanics.

With W. Heitler in the laboratory, we soon grasped that wave mechanics was not so simple, but Mott's lectures were a memorable introduction to the new theoretical physics. Later we found that he had given this course of lectures a few years earlier, when he was a lecturer in theoretical physics at the University of Manchester. The small book of those lectures, *An Outline of Wave Mechanics*,[1] was a classic of that time and a tremendous help to those of us, up to then immersed in classical theory, who had to face the examination papers only months ahead.

In his introduction to that book, Mott wrote that in Manchester his aim had been to give 'an account of the methods of the New Quantum Theory that should be intelligible to the advanced student of experimental physics, and to the research workers.' That, was the attitude which made Mott's presence in the Bristol physics department so important. A. M. Tyndall, the director of the laboratory, had built up an experimental laboratory of diverse researchers and had sought Rutherford's advice about a successor to Lennard-Jones. Later in life Mott[2] revealed that he was reluctant to leave Cambridge, but many of us are thankful that the Rutherford–Tyndall axis persuaded him to do so. Mott might well have spent his time in working with the famous figures that were

---

* J. E. Lennard-Jones moved to the Plummer Chair of Theoretical Chemistry in Cambridge and for a brief period before he died in 1954 was the first principal of the University College of North Staffordshire, Keele. For an account of his distinguished career see Mott, N. F. (1955) *Biog. Mem. Roy. Soc.*, **1**, 175.

soon coming to the laboratory – Heitler, Bethe, Zener, Gurney and many others created a vibrant atmosphere in the mid-1930s. But Mott was greatly interested in the experimental work of the laboratory and those of us who became research students were given enormous help.

In my own case Mott cleared away the mystery of my observations. I had been given the task of investigating the abnormal electrical resistivity of thin films of the alkali metals deposited on a glass surface in high vacua. There was a problem from earlier observations that even films as thick as a thousand atomic layers still had resistivities several times greater than of the bulk metal. Working under very pure conditions, my own films behaved more reasonably. Although stable from single atomic layers, the resistivity remained significantly greater than for the bulk metal until the films were more than about 40 angstroms thick. I could not devise an explanation and went to Mott. He looked at my results and in a few minutes showed me how to develop a quantum mechanical explanation, instead of a classical one, in terms of the shortening of the mean free path of the electrons through collisions with the film boundaries.[3]

Soon I had a far more important reason to be grateful to Mott – and it had nothing to do with theoretical physics. In the summer of 1936 he and Tyndall thought it was time for me to move on, and arranged for me to be interviewed for a post as Assistant Lecturer in Physics in the University of Manchester. Reluctantly I journeyed to Manchester and was interviewed by

H. H. Wills Laboratory, University of Bristol.

15

Staff of the Physics Department, University of Bristol (1935): (front row, left to right) W. Sucksmith, H. W. B. Skinner, G. I. Harper, L. C. Jackson, N. F. Mott, A. M. Tyndall, S. H. Piper, I. Williams, H. H. Potter, W. Heitler; (middle row) Mr Huntley, S. E. Williams, K. Fuchs, H. Jones, E. T. S. Appleyard, Mr Baber, V. E. Cosslett, C. F. Powell, J. H. Burrow, K. Worsnop, A. F. Pearce; (back row) R. W. Gurney, L. Frank, H. Worthy, W. R. Harper, Mr Mercer, A. C. B. Lovell, N. Thompson.

Bragg and his forbidding staff. Glad to return to the ambience of the Bristol laboratory, I soon encountered Mott, when the following conversation took place.

Mott: I hear from Bragg that you have been offered the job.
Lovell: I'm not going to accept. It was dark and wet and I should have been playing cricket here.
Mott: Don't be a fool, Lovell. Don't you realise that Manchester is the home of the Hallé and the *Manchester Guardian*, and that Old Trafford is near the lab.

Perhaps Mott had recalled his own reluctance to accept the Bristol offer, but I remember that conversation more than sixty years later, still near Old Trafford and the Hallé.

References

1. MOTT, N. F. (1930) *An Outline of Wave Mechanics*. Cambridge: Cambridge University Press.
2. MOTT, N. F. (1986) *A Life in Science*. London: Taylor & Francis, p. 44.
3. LOVELL, A. C. B. (1936) 'The electrical conductivity of thin metallic films'. *Proc. R. Soc. A*, **157**, 311.

PROFESSOR SIR BERNARD LOVELL, OBE, FRS, was born in Gloucestershire in 1913 and educated at Kingswood School, Bristol. He obtained BSc and PhD degrees from the University of Bristol and, after three years at the University of Manchester, he worked from 1939–1945 on the development of airborne centimetric radar at TRE. He returned to Manchester at the end of the war, and in 1951 was appointed to a chair in radioastronomy. He was knighted in 1958 for his services to science. Among his many achievements, he was responsible for the construction of the Jodrell Bank radio telescope.

Early Days

David Shoenberg

As one of the older contributors to these reminiscences of Nevill Mott, it is perhaps appropriate that I should concentrate on my memories of his younger days. By the time I came as an undergraduate to Cambridge in 1929, Nevill had already acquired a considerable reputation for his applications of the relatively new wave mechanics to problems in nuclear physics. His most striking

Nevill Mott with Tyndall, Bristol 1935.

result was to predict that the scattering of α-particles by helium should depart from the famous Rutherford formula, which had proved so successful for scattering by other nuclei. The significant feature of scattering by helium that made it anomalous was that the scattered nucleus (the α-particle) was identical with that of the scatterer. Chadwick's experimental confirmation of Nevill's prediction caused quite a sensation. After a year in Manchester, Nevill returned to Cambridge in 1930 as a fellow of Caius and gave a course of lectures in the Cavendish in which he emphasised the physical meaning of how wave mechanics works. I am ashamed to say that I cannot remember much about the lectures, which I attended in 1931, except that they were held in the middle of the afternoon at a time when it is easy to succumb to sleepiness. I had very recently converted from mathematics to physics and his approach of emphasising the physical significance rather than the mathematical formulation did not then mean much to me. Not long afterwards he published his first book, *Outline of Wave Mechanics*, essentially the substance of the lecture course, and I was much better able to understand what it was all about.

When Nevill moved to Bristol in 1933, as Professor of Theoretical Physics, he began to take an active interest in several new fields which were already being studied there. Prominent among them were various aspects of the behaviour of metals, and characteristically Nevill organised a conference in 1935 to which he invited everyone active in the general area from all over the world. These included not only well-known figures such as Casimir, C. G. Darwin, Gerlach, Simon and Stoner, but lesser lights such as myself, who had only recently started research. At that time I was working on the peculiar magnetic properties of bismuth, to which I had been introduced by Kapitza when he supervised my PhD. I had already had some contact with Norman Thompson in Bristol; he was measuring the resistive properties of bismuth and its dilute alloys. And Willie Sucksmith had advised me how to construct a copy of the magnetometer named after him. I was flattered to be asked to present my latest results at such a prestigious meeting and I still remember with pleasure the very pleasant and informal atmosphere created by Nevill, which facilitated exchange of ideas between people of widely different experience and achievement. One of my photographs, illustrated here, captures Nevill's cheerful and friendly manner which made the conference such a memorable success. This was effectively my first personal contact with Nevill and resulted in some helpful discussions about the interpretation of my experimental results. For me, one very useful consequence came a year later when Nevill lent me a proof copy of the book on metals and alloys he had just written in collaboration with Harry Jones. I took it with me on a trip to Eastern Europe in 1936 and I remember reading it by the lakeside of Zell-am-See in Austria on my way home from Russia. The book has provided a valuable background to much of my work ever since.

During the war, Nevill occasionally visited Cambridge, where I was still based, and would sometimes drop into my office for a gossipy chat. I cannot now recall what we talked about – mostly I think current affairs and physics in the Soviet Union – but I do remember his friendliness and complete lack of 'side'. There are many anecdotes about Nevill's absent-mindedness, most of them probably apocryphal, but I should like to mention one which concerned his wartime activities. He was working at an RAF secret establishment in Stanmore, near London, and the story goes that the guard at the gate noticed Nevill walking up and down the drive just inside, evidently in deep thought. After a while he came up to the guard and asked him, 'Was I coming in or going out?' The reply was that he was coming in, 'Oh, good,' said Nevill, 'I must have had my lunch.'

Our relations became a good deal closer when he returned to Cambridge in 1953 as Cavendish Professor, and he provided helpful support and encouragement for the work of Brian Pippard and myself in the Mond Laboratory on superconductivity and the electronic structure of metals. I also particularly remember the warm hospitality of Ruth and Nevill at their home in Sedley Taylor Road and later, after he became Master of Caius, at the Lodge. In more recent times he often lunched in college and I enjoyed the occasions when I happened to sit by him and once again we chatted about all sorts of things. To conclude, let me mention some particular examples of Nevill's kindnesses to me during his time in postwar Cambridge. He had many more invitations for functions abroad than he could accept, and on two occasions he asked me to deputise for him. He knew of my interest in Russian affairs and in 1964 I had a very interesting trip going in his place with a Labour Party delegation to Russia to study their educational methods. Another occasion was almost entirely of a social nature and involved attending the formal opening of a new *haute école* in Paris. The hospitality, both gastronomic and alcoholic, was fabulous and it was a wonderful perk. Finally, I was grateful for valuable support from Nevill in college in 1960. I had at that time made something of a breakthrough in my research on Fermi surfaces but found it more and more difficult to combine intensive work in the laboratory with teaching duties in college. When I discussed the position with Nevill and asked if I could be relieved of my teaching duties, he was sympathetic and succeeded in persuading the rather hard-nosed College Council to agree to my request without depriving me of my Fellowship.

Other contributors will no doubt enlarge on the great scientific significance of Nevill's contributions to so many aspects of solid-state physics, following his early work on nuclear and atomic physics. However, I hope that my own recollections will have conveyed something not only of his kindness and friendliness, which have meant so much to me over sixty years or so, but also of his scientific style and versatility, which have made him such a leader in the world of physics for even longer.

DAVID SHOENBERG is Emeritus Professor of Physics, University of Cambridge. He was born in St Petersburg, Russia, in 1911 but since 1914 has spent most of his life in England. He was educated at Latymer Upper School in London and entered Trinity

College, Cambridge, as a mathematics scholar in 1929, but changed to physics after his first year. He started research under Peter Kapitza in the newly established Royal Society Mond Laboratory in 1933. In 1947 he was elected a fellow of Gonville and Caius College and also became head of the Mond Laboratory, which in 1972 moved to West Cambridge with the Cavendish Laboratory. His research has been mainly on superconductivity and the electronic structure of metals.

A Miraculous Escape

Lilly Gill

Shortly after the Munich Crisis, in late 1938, when Chamberlain narrowly averted Hitler's invasion of Czechoslovakia, my father was put into contact with Nevill and Ruth Mott through a chance meeting in London between Ruth's sister, Mary Horder, and a Quaker lady, Tessa Rowntree, who was helping Jews in Prague to flee from Nazi aggression. Tessa Rowntree was desperately trying to help my father, a professor of music who had fled from Berlin to Prague, to go to Toronto, Canada, where he had been offered a post at the Academy of Music. The first stepping-stone towards him going there was to find someone willing to act as guardian to his two youngest children. Mary Horder suggested that her sister and brother-in-law might be very interested in helping such a family. Imagine our joy when my parents received a letter from the Motts in England spontaneously offering to care for myself, age 16, and my youngest sister, age 11, until such time as my father was able to secure a better future for us. Many letters were exchanged between Nevill and Ruth and my parents during those anxious weeks before Hitler's invasion of Prague by the Nazi troops in mid March 1939.

After a terrifying escape journey by almost the last train to leave the city, and finally a boat crossing from the Hook of Holland to Harwich, my sister and I arrived in London. Nevill and Ruth came to meet us at Fenchurch Street station. Their kindness and generosity seemed unbelievable and it was so comforting to talk to them in German, as we had very little knowledge of English then. After a few days stay at Two Oaks, Ruth's family home in the heart of Sussex, near Haywards Heath, we went to live with Ruth and Nevill at their flat at Caledonia Place, Clifton, Bristol. This flat, however, became a bit small for us all and I can remember the excitement and fun of looking at a much bigger one. Nevill liked the architectural aspect of the beautiful crescent-shaped group of houses, Princess Buildings. I believe they were Edwardian and rather grand and tall, right on the edge of Clifton Gorge, overlooking the river deep down below and flanked by the Clifton Suspension Bridge. We were to live in the two top stories and a roof garden came with it! At weekends Ruth and Nevill would take us for delightful outings into the country in their lovely old car, Emily, an Austin 10 with a canvas roof and Perspex windows. Sometimes these outings were to places nearby, e.g. Clifton Downs, where we played a lively game of French cricket and had a picnic, sometimes further afield to the Mendips. Nevill was always wonderfully kind and fatherly towards us both. Perhaps better than words, the little pencil sketch which I

Lilly Gill's wedding at St Agnes' Church, Liverpool (1949): the bride was given away by Nevill, second from right.

sent to my mother (and reproduced here) illustrates the happiness surrounding our day-to-day life with the Motts: Nevill giving my little sister a piggyback to help her up the rather steep hillside, I believe somewhere on the foothills of the Mendips. Nevill, a very keen hill-climber himself, insisted on purchasing walking sticks for each of us, as well as strong waterproof shoes; he explained that the correct gear was most important.

We were sent to different schools at first. For me they chose St Brandon's Clergy Daughters' School in the heart of Bristol and my sister attended Duncan House School on the Clifton Downs. Ruth was a very accomplished harpsichord player and Nevill loved playing the clarinet, so we were given plenty of encouragement and opportunity to develop our musical potential. I remember, fondly and vividly, playing the piano accompaniment of Mozart's Clarinet Concerto with Nevill – we had such fun. My sister took up recorder playing; Ruth and Nevill were both very keen on recorders and had a very fine collection of Karl Dolmetsch's instruments of various sizes. Thus the spring and summer passed very pleasantly, but war was imminent. St Brandon's School was evacuated to Wells, Somerset, and we became boarders. Not unlike the girls of St Trinian's, the girls of St Brandon's left their mark when we were sharing the Bishop's Palace with the poor bishop, who died not long after the invasion by the daughters of the clergy! Ruth and Nevill came to visit us often at weekends, taking us out to lunch or tea. Nevill had a sweet tooth and loved the gorgeous cakes and scones at some of the farmhouse tea places.

In the Master's Garden, Caius College (1960): Nevill and Ruth entertain Lilly Gill (second from left) with her mother and husband.

Lilly's drawing of Nevill giving her sister Ilse a piggyback in the foothills of the Mendips; they are followed by Ruth and Lilly.

Nevill would often talk to me and explain about beautiful architecture, and he made me so much more aware of the grandeur of old buildings like lovely Wells Cathedral, the Bishop's Palace, etc. Quite soon after St Brandon's School had been evacuated, Ruth joined the teaching staff, teaching classics, Latin and Greek. She found suitable accommodation at a small nearby farmhouse and Nevill used to come and stay with her at weekends. But the war was not going well for England in the early stages, and Bristol was targeted for heavy bombing. Ruth and Nevill's home was under constant threat and the anxiety became intolerable. Classed as an enemy alien, I was nearly sent to the Isle of Man and was forbidden to return to school in Wells, which had become a protected area. Fortunately, Nevill's sister Joan, who with her husband ran a school in Lancashire, kindly let me spend the 1940 summer holidays with her there. All Nevill's efforts to allow me to return to school and continue my education failed; he wrote pressing letters to the Home Office on my behalf, but to no avail. Nevill and Ruth then helped me to make the right decision for a career. It was to be teaching and most generously they sent me to the Charlotte Mason Teacher Training College in Ambleside for a two-year training course. Meanwhile, Ruth and Nevill had become parents to a baby daughter, Elizabeth, so when my teaching training finished, I went to them in Sussex as a mother's help to Ruth, who was by now expecting her second child, Alice. At the christening a few weeks after Alice's arrival, I became her godmother.

While staying at Nevill's sister's school, I became friendly with a young man, Michael Gill, who purely coincidentally was starting his Royal Naval Volunteer Reserve (RNVR) training as a midshipman near Hove in Sussex. We saw quite a lot of each other and became unofficially engaged. I shall always remember Nevill's kindness when my future father-in-law tried to put an end to our attachment. He wrote to Nevill, urging him to intervene immediately and pointing out that his son's naval career might be at risk by associating with an 'enemy alien'. Nevill wrote back forthwith, ridiculing such strong measures. Thereby he put an end to the whole unfortunate and unpleasant episode and my heart was filled with enormous gratitude towards him for the prompt action he had taken. The young naval officer became my husband on 4 January 1949 and we were married in Liverpool, his home town. Nevill, pictured in the photograph, attended the wedding as guardian/father and gave the bride away. In October 1950 our first child, John Nicholas, was born and Nevill became his godfather. We often went to see Ruth and Nevill. Nevill had become the Head of the Cavendish Laboratory in Cambridge, and Ruth and he were living at the Master's Lodge of Gonville and Caius College. My mother had left Prague for Sweden; she came to England several times and visited the Motts and us, as the photograph shows. I believe Nevill liked my husband, Michael, very much and they had long and interesting discussions when we visited. Tragically, my beloved husband died in October 1965 aged 40; Ruth and Nevill shared my grief with great compassion and understanding, and they invited me to spend Christmas with them, along with my three boys.

A year or two later, when Nevill received a knighthood from the Queen, Ruth and Nevill visited us on the way to Buckingham Palace, and we certainly felt very proud to think we shared part of that great occasion with them. During the last two or three years I saw Nevill quite frequently, when I went to their home in Aspley Guise to be with Ruth as a companion while Nevill was in Cambridge for a few days at a time. Not only was this a golden opportunity to do something in return for so much kindness that I had received in the past, but it also proved a providential occasion to see Nevill and to talk at length about many things concerning the past; things he had really never known before. When I showed Nevill the photograph of a painting, painted in 1891 of my father playing the piano at the court of Franz Joseph, the last emperor of Austria, he was truly amazed and tremendously enthusiastic about the idea that I should go to Vienna to seek the painting, and there and then insisted on helping me with the expense. I did go, last May; I visited the house where my father was born and the concert hall where he gave concerts, but alas I could not find the painting; however, I shall continue with my search. The news of Nevill's death came as a great shock. Attending his funeral was very sad but it was extraordinarily uplifting and inspiring as well. It is so hard putting things we feel very deeply into words. To me, Nevill was a very great man, of immense intellect, but also an unbelievably kind, charismatic and caring human being whose memory I greatly treasure.

Der „kleine" Spielmann bei Hofe.

LILLY GILL was 17 when she came to live with the Motts in Bristol, following a perilous escape from Nazi-occupied Prague in March 1939. In 1940–41 she attended the Charlotte Mason College in Ambleside for a two-year teacher training course and began her teaching career in 1943 at St Hilda's PNEU school in Bushey. After her husband died in 1965, she dedicated herself to her sons' upbringing and education. She has artistic inclinations, mainly ceramic sculpture and drawing. Her father, Leopold Spielmann, was a child prodigy; the drawing shows him playing the piano before Emperor Franz Josef in 1891. He was a victim of the Holocaust, dying in a concentration camp in 1940.

Nevill: Family Friend

Eleanor Scott

I first met Nevill in Bristol, my home town, at the very modern house that Herbert Skinner, later a professor at Liverpool, had built for his glamorous German wife, Erna. I was very shy and completely overawed by the high intellectual content of the conversation, which seemed to be all about the latest scientific ideas, so I took refuge with Erna, who like me was no scientist.

Later on, through Ruth, who like me was teaching classics, I got to know him better and realised that far from being like some scientists – notably Dirac, who would sit through a whole meal without saying a word – he was a very warm person with wide and varied interests. In their own flat, overlooking the Avon Gorge with its huge fluctuating tides, one could be sure of lively conversation on a variety of topics. He had bicycled or walked all over Somerset, even as far as Venice, and was keenly interested in the architecture of the many beautiful churches. He shared Ruth's interest in numismatics; great was the excitement when they got an authentic shekel. He was well informed about garden plants, though I never saw him digging when eventually they had a big garden.

They loved Bristol; at that time ships came up the Avon right to the centre of the city. There were good concerts and besides a repertory theatre, the Theatre Royal, where people still threw nuts at the actors if they didn't like them! Most of that has gone now, but nothing can take away the splendid views of the Mendip Hills. Moreover, there was a good mixture of town and gown; sherry and tobacco were the predominant industries. Links between university and industry had always interested Nevill and later he sat on various committees to integrate them.

Then came the war, and he visited us at Mudeford near Christchurch, where my husband was working on radar. Together we watched the first dog-fight in the air. A huge fleet of German bombers came over, unmolested. Then suddenly from all directions came British fighters; a number of enemy planes were shot down and the rest retreated. Radar worked! It was most exciting, and somehow one didn't think of the people inside – till Nevill said, 'I hope the poor devils got out safely and have been taken prisoner.'

In the early days of the war, to their great joy, their first child Elizabeth (Libby) was born, after nearly eleven years of marriage. Two years later, in 1941, came another daughter, Alice. And following Alice, in the same year, our own first child was born. We had a son and Nevill was one of his godfathers.

Nevill was now a family man and he much enjoyed that role, filling it admirably. But after a while there came a great shock: Libby was diagnosed as badly and incurably handicapped. While in Bristol they could just manage, as after the war they had a large house near the Lab and were able to get live-in help. But in Cambridge things got much worse; now in her early teens the child became more exacting and could never be left alone in the house. So after much heart-searching it was decided that she needed residential care. An establishment was found, run by Roman Catholic nuns. They were both absolutely devoted to her. With tears in her eyes Ruth said, 'But what will happen to her when we are gone?' She was given the best medical treatment available at the time, and Nevill overcame his fears when he saw how happy she was; the annual convent fete took precedence over all other engagements.

Much though they missed Libby, it did give Ruth ample free time and Nevill helped her to form a society of Cavendish wives. They met once a month for coffee, a chat and a lecture. This was welcomed as a chance to get to know other wives, sometimes difficult in a fragmented and rather tribal society. He also supported Ruth in her musical interests, and got a special car which would take her clavichord when she played in concerts.

He felt immense pride and pleasure when Alice got one of the first coveted places at the Inominate Third Foundation (the other two being Newnham and Girton) which became New Hall. And I remember travelling up to London with him; he had taken a day off to dissuade her from joining a party of VSO volunteers to Cambodia, dangerously near the war in Vietnam.

He extended his warm feeling to our family, too – and I imagine to his other colleagues. He wanted his godson Martin to go to Caius (which he hoped to make into a Cavendish College). But Martin was offered a postmastership at Merton College, so he decided to go to Oxford. With some trepidation I broke the news to Nevill, who had always been remarkably kind. For instance, when our youngest son had a nasty accident, putting his arm through the solid glass panes in the front door, he was the first person to come round and offer help. And when my husband had shingles, Nevill was most sympathetic. We never advertised these disasters, but he seemed to get wind of them.

When he moved to Milton Keynes, he delighted in his three grandchildren and kept a trunk full of his honorary degree hats and gowns for them to dress up in. His grandson, Edmund, got top marks for physics and a prize at Imperial College; fortunately the news broke just before Nevill died, and it was satisfying for him to know the torch was being carried.

Typically, when he got his knighthood he said it meant little to him apart from the satisfaction it gave to his many friends. During a party in our house, a cheeky grandson asked him, 'What did you get knighted for?' He replied, 'It goes automatically with the job.' He did, however, admit that he was rather thrilled to get the Nobel prize. He loved entertaining and we spent happy evenings in the magnificent eighteenth-century rooms in Caius, and also in their home in Sedley Taylor Road.

After his retirement, he used to come and talk to me about religion; he had read far more than most theological students, in foreign languages too.

(Incidentally, how many physicists could lecture in French and read German with ease as he could and did?) I think he used me more as a sounding-board for his own views; I was a good listener and would make the occasional interpolation and objection, but nothing more. For instance, although I prefer the 1662 prayer book, I realise that it is meaningless to many young people, and one has to modernise.

Eventually Ruth became too frail to live at home. The head of the excellent Cambridge establishment which looks after her said to me, 'It's lovely to see them together, it's only a shame they can't be near each other all the time.' I cannot overemphasise the part Nevill's sister, Joan, played throughout all this time. She was a constant support in all the crises, would step in and do anything, and latterly acted as chauffeur, providing beds in Cambridge for anybody from the family who needed them.

Nevill loved the New Cavendish – even the food! And it must have been a constant satisfaction and crown to his long career to see his name in bold letters on the Mott Building.

MRS E. SCOTT (née Dobson) graduated in 1933 in the Classical Tripos at Cambridge. She taught at St Alban's High School, St Paul's Girls School and Homerton College, Cambridge. In 1939 she married J. M. C. Scott, a physicist. She now teaches Modern Greek.

War Years

Bristol 1938: Memories of a Research Student

Frank Nabarro

My connection with Nevill Mott began in 1937, when I took my degree in physics at Oxford; he was the external examiner. My performance in the practicals had already marked me out as a theoretician. He agreed to take me as a student, but advised that I should first take a degree in mathematics. This may have related to his own career. He relates in his autobiography that he condensed his study of mathematics in Cambridge into less than the normal number of years, and ever afterwards felt that there were gaps in his arsenal of mathematical weapons which sometimes may have left him at a disadvantage. One of these gaps must have been the theory of groups. I don't think he ever used this theory, nor were his disciples in Bristol ever encouraged to learn it. These accidents have long consequences. I was responsible for the Department of Physics in the University of the Witwatersrand, Johannesburg, for many years; very few of my colleagues or their students ever made use of this theory.

Nevill and I published our first paper together in 1940, in the report of the first of the Bristol Conferences, which he organized. He was already concentrating on the style of physics for which he became renowned – a combination of deep insight with a minimum of mathematics. I think people tend to forget how powerful a mathematician he was. His early papers on the scattering of a particle by an electric charge, first using the Schrödinger equation, then using the Dirac equation, show this very clearly. But the mastery remained in reserve. I was his assistant at various times during the Second World War. At one time I was concerned with the question of how far the operator of a searchlight should stand from his projector. If he was too near, he was dazzled by the light scattered back from particles in the air; if he was too far away, he had trouble steering the beam. The problem involved the evaluation of an unpleasant integral. I could not do it. I consulted J. M. Whittaker, an FRS and professor of mathematics, who solved it for me, with some difficulty. Mott did it 'on the back of an envelope.' Hans Bethe once wrote that, 'There are two types of genius. Ordinary geniuses do great things. ... Then there are magicians.' In that framework, I felt that Mott was an ordinary genius. His mind worked like other people's, only better. But this allowed him to be a human being, whereas magicians often are not.

By no means desk-bound, we visited anti-aircraft research establishments and operational sites, usually on bitter winter days. At that time, Nevill was very slim and felt the cold badly. He would wear three pullovers, first a blue, then a grey, then another grey. At night we would sleep in the local inn. Nevill would dress for dinner, first two grey pullovers, then the blue.

The obituary notice in the *Times* tells of a later period, when he was consulting at Harwell. He returned to Bristol, forgetting that he had already moved to Cambridge. The story current among theoretical physicists was, naturally, more elaborate. He was still living in Bristol. He drove up to Didcot with Ruth, and put her on the train for Oxford, where she went shopping and visiting friends. After his consultation, he drove back to Bristol, only to discover that both Ruth and the keys of the house were in Oxford.

I arranged to go to Cambridge on sabbatical from Johannesburg. Nevill had promised to support my visit with a considerable sum of money, I think £300. When I arrived, he had no memory of his promise. Fortunately, I had brought his letter. The problem was that he had given the money to someone else. However, he found the alternative solution that he could pay me almost as much if I agreed to give a course of lectures on 'anything at all'. In bigger matters he was methodical and effective. At the end of the war, he 'arranged' for me to have an award to work with him in Bristol. After four years, he 'arranged' that I should go to Birmingham for contact with Alan Cottrell.

During one of my later visits to Cambridge, my wife and I had Nevill to lunch along with our son, Jonathan. Jonathan could not get over the fact that this Nobel laureate arrived with his trousers held up by a piece of string.

On the personal side, Nevill could be very direct. After one of Alan Cottrell's stimulating visits to Cambridge, Nevill remarked that he looked more like a farmer than a physicist. Then he added, 'Come to think of it, so did Rutherford.' Mott shared some of Rutherford's scorn for practitioners of other sciences. He used to say, 'Rutherford got the Nobel prize for chemistry the same year the German Kaiser got the Nobel prize for peace.' And when I reintroduced him to my wife-to-be, Margaret Dalziel, whom he already knew slightly from our work together in the Army Operational Research Group, he said to her, 'Do you know what you are letting yourself in for?' (She said yes.) Margaret's other abiding memory of that time is of the live performances of chamber music in Royal Fort House. Nevill also had a strong respect for the elders of science. He revered G. I. Taylor. After G. P. Thomson had handed over the executive editorship of *Phil. Mag.* to Mott, Thomson sent in a paper which Mott felt was in serious need of revision. Mott consulted Tyndall, and they agonized over the situation. The solution was, 'Well, after all, G. P. Thomson *is* G. P. Thomson.'

His sensitivity to the feelings of others appeared in a number of ways. After the war, he assembled a very bright group of PhD students and post-docs, many of them from countries which had been under German occupation. Mott sent me to visit Becker's group in Göttingen. There I met two very impressive young postdocs, one male, one female. Mott had only enough money to invite one of them over. After we had discussed their possible contributions to the work of the group, Mott said, 'Let's have the girl. It will be less embarrassing to the people from occupied countries.'

I am left with one problem. The head of the Army Operational Research Group during much of the war was Basil Schonland. Mott records in his autobiography that they could not work together. I worked under both. They were very different, but both were totally honest, and I profited from both experiences. I never understood what came between them. Oddly enough, they were awarded knighthoods at about the same time, and both became Honorary Fellows of Caius College, Cambridge.

PROFESSOR F. R. N. NABARRO, MBE, FRS, was born in London in 1916. He was educated at Oxford, Bristol and Birmingham. He is Emeritus Professor of Physics at the University of the Witwatersrand, Johannesburg, and a consultant and Fellow of CSIR, Pretoria, South Africa.

Fort Halstead: Superintendent of Theoretical Research in Armaments

Ian Sneddon

As one of the group who had worked under the direction of Professor J. E. Lennard-Jones at the Mathematical Laboratory, Cambridge, I moved to the Armament Research and Design Department at Fort Halstead when he became its director early in 1944. I was assigned to the branch of Theoretical Research in Armaments of which Professor N. F. Mott was the superintendent. Having as a physics student at Glasgow University read his *An Outline of Wave Mechanics* and dipped into Mott and Massey, and Mott and Jones, I had expected him to be a rather formal and remote person. What I didn't know was that these books had been written before his thirtieth birthday, and at the age of 29 he had been appointed a professor of theoretical physics – a rare distinction in those days! So I was surprised to be warmly welcomed by a young man – tall, slim and blond, more like the head boy of a public school than my idea of a professor.

The staff of the branch had a few senior scientists: J. W. Maccoll (deputy superintendent) and E. Dearden were from the permanent civil service; the only establishment academic was L. Howarth, who had come from a lecture-ship in mathematics at Cambridge. The remainder (from the Ordnance Board, or recruited by Lennard-Jones to the group at Cambridge) were recent PhDs (F. Booth, J. Corner, A. F. Devonshire and E. H. Lee) and wartime graduates (N. Clerk, J. Codd, E. Hicks, R. Hill, C. F. Illingworth, D. C. Pack, K. Thornhill, S. J. Tupper, J. H. Wilkinson and M. Wood).

We were organised into groups on fluid mechanics, solid mechanics and internal ballistics. At Cambridge I had worked on problems of armour penetration under W. R. Dean and J. W. Harding and had contributed to a phenomenological theory of armour penetration, so I was an obvious candidate to join the 'solids' group under Mott. Of course, we did not just work to a research programme; we sometimes had to drop everything and provide a solution to a day-to-day problem. Such solutions ranged from giving an instant reply on the telephone to a service officer requiring information, to working out by a numerical procedure the inertia tensor of an odd-shaped projectile. But, on the whole, the bulk of an individual's time was spent on a long-term project.

I remember that the first discussion I had with Mott was on the strength of steel used in the manufacture of tank armour plate. It was a good order of magnitude less than K. Fuchs had predicted in a paper published just before the war. The discrepancy between the theory and practice was disturbing. After a short delay (of some days) Mott called me into his office again and told me that tank armour plate had been manufactured, not by Woolwich Arsenal but by the Ideal Boiler company! He then speculated that when the bulk steel was produced it included spherical bubbles of gas, flattened during the roll process into disc-shaped cracks. Mott suggested the name 'penny-shaped cracks.' He then gave me a long-term project to study the effect of these cracks on the mechanical properties of steel plates.

It did not take me long to solve the canonical problem: the case where a single crack is situated at the centre of a plate which is symmetrically loaded. I suddenly saw how it was closely related to the penetration problem that J. W. Harding and I had tackled at Cambridge. The basic analysis was rather complicated and had only been published by E. C. Titchmarsh in a book that appeared in 1937. When I had written out the solution, I took it to Mott and asked him if we should have it produced as a departmental report (classified as secret and therefore of limited circulation). His response was immediate: 'Yes! But print tens of thousands of copies and have the RAF drop them over Germany; then the Nazis will realise they can't defeat a country that fights a war that way!' Then he gave me a valuable lesson on how to present a report to a wartime Defence Committee whose membership consisted of civilian scientists and engineers along with military officers (usually majors) only some of whom had a BSc degree. Most of the members hadn't read the papers before the meeting, and when you started speaking they would be looking at the introduction then they'd turn to the back to look at the 'pictures' (in those days all the diagrams were collected at the end of the report). So Mott said, 'State your physical ideas clearly and your final results in one or two simple formulae in the introduction, depending on graphical methods to get more sophisticated results across.' Though he attended meetings at which one of his departmental reports was being presented, he might contribute to the discussion but he left the actual presentation to the individual author, and did not mention any part he had played in the production.

After I had written my report, Mott revealed that he had given the crack problem to R. A. Sack, a young refugee, who on release from an internment camp was doing some work in the H. H. Wills Physical Laboratory at Bristol and had arranged for him to visit the fort. I was used to this kind of situation; I knew that any numerical work done on the differential analyser at Cambridge was repeated on the similar machine at Manchester. It was only if the final results disagreed that there was a slight panic! Fortunately, my result agreed with Sack's, though we had used entirely different ways to obtain it; the outcome was that we each published a paper on the subject after the war was over.

The long-term problem that E. H. Lee (whose PhD was in engineering) and I were given was to investigate gun jump. Why did a projectile aimed at the centre of a target seldom hit it? Since the design of a gun is very complicated – I wished we still fired spherical shot from cylindrical barrels – we had first of all to devise a model. It was here that Mott's physical intuition was invaluable in isolating the essential geometrical features and the significant parameters.

I was conscious that Mott was also closely in touch with the plasticity group and did individual work on fragmentation, which I knew had interested him since his spell at Bomber Command.

What I remember most vividly was the end of Mott's period as superintendent. A few of the staff had academic or civil service posts to return to once the war was over, but the majority had to find new careers. Some wanted to be made permanent scientific civil servants, others to begin a career in the academic world (scientific or administrative). He spoke with each of us and discussed our aspirations; in the end he spent a great deal of effort to ensure that as many of us as possible were well placed.

One other thing I remember about those last days. Knowing that his prewar work had been entirely on atomic physics, the press assumed that he had been involved in the design of the atomic bomb. When a young woman reporter asked him directly if he had, he replied, 'No, my dear, I only worked on those weapons that God always intended us to have.'

PROFESSOR I. N. SNEDDON graduated from Glasgow University in 1940 and Cambridge University in 1942. After was service he returned to Glasgow as a lecturer in theoretical physics. In 1950 he was appointed Professor of Mathematics at University College of North Staffordshire (now Keele University), moving back to a similar appointment at Glasgow in 1956. He took 'premature' retirement in 1982 but is still active in mathematics.

My Professional Association with Sir Nevill Mott

Jack Allen

Aside from casual contacts with Nevill, I had only two serious professional meetings with him. One was in the Second World War when I was working on the elastic behaviour of shells in going up the barrel of a gun, mostly the 4.5-inch AA gun. The other was at a meeting of the General Assembly of the International Union of Pure and Applied Physics (IUPAP) in Rome in 1957. Mott was chairman and can be seen in the front row of the photograph.

My work on guns started when David Shoenberg and I were asked by Sir John Cockcroft to design an electronic variable time fuse, containing hot-filament glass vacuum valves to replace the highly ineffective clockwork fuses then in use in the Blitz. They had to be set before loading, so precise aiming was impossible. The electronic fuse could be loaded and its timing set at the instant of firing, with the aid of a pick-up coil mounted on the gun muzzle. David did the electrics and I did the mechanics of assembly and mounting the fuse parts, so they survived the shock of firing. The scientists at Woolwich assured me that the shells could be safely treated as mathematically rigid bodies. How wrong they were, but understandably so, since they were applied mathematicians.

I soon found, after initial disaster, that the shell must be vibrating violently in the barrel. I devised a gauge that would record the magnitude of a force which lasted less than one millisecond. This showed that the shell hit the barrel of the gun before engaging the rifling with a force as high as $20\,000g$. This was a transverse force called side-slap. Then, by Poisson's ratio, it also produced longitudinal or axial forces of comparable magnitude, which I also recorded. During this work, I carefully examined the copper driving band on a recovered shell to see if the surface had melted. I could find no sign of this. I encountered Nevill and told him about this. He replied that, in friction, as the metal surface got close to the melting point it softened and the frictional heating declined so that the melting point was never quite reached. This was an eye-opener. Then I thought that this might affect the axial vibration. Since the driving band is far back from the centre of mass, then in axial vibration the value of the coefficient of friction will vary with the phase of vibration. If the shell is lengthening, the rubbing velocity is lower, so friction is higher; if the shell is shortening, the rubbing velocity is higher, so friction is lower. This is equivalent to positive feedback, so any incipient axial vibration will speedily increase in amplitude.

Jack Allen (arrowed) and Nevill (front row) at the IUPAP General Assembly in Rome (1957).

I designed a gauge to measure this and was delighted to see a record of four or five vibrations of rapidly increasing amplitude before the shell reached the muzzle. This was done in a cold gun that had not been fired since the previous day, and it was in February. I arranged another trial a few days later, again in a cold gun, and this time with a screen about 100 feet square and about 200 yards in front of the gun. On firing, two holes appeared in the screen, one about 4.5 inches in diameter, made by the shell, and another a few feet away and smaller, made by the fuse. The axial vibration had been violent enough to shear the brass threads of the fuse which held it to the shell. This was great, since it solved a problem in warfare that had troubled armies for a hundred years. In the artillery barrage that preceded the attack, the opening rounds, from cold guns, either misfired or went astray, and gave away the gun positions to the enemy. I would never have thought of this if it had not been for Nevill.

At the triennial meeting of the General Assembly of IUPAP in 1957, held in Rome, I was one of the four or five UK delegates and Nevill Mott was chairman. I enclose a picture of us all. The meeting turned out to be of crucial importance. The Cold War was building up, McCarthyism was still strong in America and there was the growing problem of Chairman Mao in China. There was a motion, possibly proposed by the Americans and certainly strongly supported by them, that physicists from communist countries should not be allowed to attend IUPAP supported conferences, nor should anyone who worked in armament research establishments. The British group was strongly opposed to any exclusion, and we found many supporters from other countries. We also checked with Mott and found that he was on our side. We therefore proposed a counter-motion that any bona fide physicist from any country or occupation should have a right to attend an IUPAP-sponsored conferences or meetings. The debate was lively but Nevill maintained firm control. On a vote our motion won and became IUPAP policy. It was a good meeting and we were lucky to have Nevill in the chair.

This had a sequel in the run-up to the Low Temperature (LT) Conference in Kyoto in Japan in 1970. John Bardeen and I were both members of the Low Temperature Commission of IUPAP. John contacted me and said the Japanese were refusing to allow physicists from the Office of Naval Research (ONR) in Washington to attend. But John said that since he was American he could not actively protest, because he might be charged with racism. Would I help? I wrote to my friend Sugawara, an LT man in Japan; he was going to be chairman at the LT conference, so I reminded him of the IUPAP ruling and warned that if the Japanese persisted, I would be forced to cancel IUPAP sponsorship and finance. Kyoto could convene a conference but it would be a purely Japanese affair. Sugawara protested to me but I said I had IUPAP backing, which I had checked, and what I had said would certainly be carried out. Sugawara relented. When I got to Kyoto, he came up to me and greeted me warmly and thanked me for what I had done. He said that he had had a furious row with the Japanese Physical Society, who had started the trouble.

At the conference reception an angry American came up to me and said he was furious with the Japanese, who had told him when he first applied to come that since he worked in the ONR they had turned his application down. I said, 'You are here, aren't you? Just be happy.' He smiled sheepishly and apologised. The conference went smoothly.

PROFESSOR J. F. ALLEN, FRS, was born in 1908 in Winnipeg. He took a BA in 1928 at the University of Manitoba and then a PhD in Toronto University in 1933, followed by a postdoc fellowship at Caltech, Pasadena. He was a research assistant in the Mond Laboratory, Cambridge, from 1935 to 1944, and a lecturer in Cambridge until 1947, when he was appointed to the Chair of Natural Philosophy at the University of St Andrews.

Some Reminiscences

Arthur Vick

I first met Nevill during 1937–39 at meetings of the Physical Society (of London) in the old physics wing of Imperial College. We chatted while we consumed tea and biscuits before the reading of papers. I was then a junior lecturer in physics at University College, London, and he was already Professor of Theoretical Physics at Bristol. I was particularly impressed by his friendliness, his enthusiasm for metal physics and by his clear exposition.

During the war I was seconded to the Ministry of Supply and became Assistant Director of Scientific Research (ADSR). After some early work on radar, Nevill succeeded P.M.S. Blackett as scientific advisor to General Sir Frederick Pile, General Officer Commanding-in-Chief of Anti-Aircraft Command (AA Command). I had frequent contact with the general, his command and the Directorates of Artillery in the Ministry and the War Office, and received copies of some of Nevill's reports on various topics. They were written rather like scientific papers for publication and left much of the interpretation of practical problems to others, but they and later reports were valuable and some continued to be influential after the war. Nevill developed a respect, as I did, for some of the professional army officers. Later the Army Operational Research Group at Petersham took over the responsibility for scientific advice to AA Command and Nevill joined the Group, where he was not happy. In 1943 he was pleased to be invited by Lennard-Jones, then Chief Superintendent of Armament Research at Woolwich Arsenal, to join him as Superintendent of Mathematics Research at Fort Halstead. At this stage I lost touch with him until after the war, but I gathered that he found his work, such as that on fragmentation, interesting and challenging. His war experiences led to a continuing interest in technology.

Nevill returned to his old post at Bristol in 1945, though he could have gone to a chair in Cambridge. Stimulated by his wartime experience, he organised summer schools on advances in solid-state physics, primarily for scientists in industry, but also for academics who had developed an interest in the subject. I attended one of these, which I much enjoyed, especially the lively discussions. Nevill was a very clear expositor. He also became a consultant to the Atomic Energy Research Establishment (AERE), Harwell, and for many years paid fairly regular visits. While I was Director of Harwell, and then Member for Research of the United Kingdom Atomic Energy Authority (UKAEA), I saw him only occasionally, since naturally he wished to spend as much time as possible with researchers, but I do know that his visits were greatly appreciated, especially by the younger scientists, who found him approachable and stimulating. Nevill was interested and pleased to find application for his physical insight.

In 1955 Nevill was President of the Modern Languages Association and from 1956 to 1958 President of the Physical Society, an unusual transition. One of his main concerns in this capacity was to take a leading part in the exploratory discussions which led to the amalgamation of the Physical Society with the Institute of Physics. I was then Vice-President, and later Honorary Secretary, of the Institute, so I was also involved. Nevill was in favour of the amalgamation for several reasons, such as the effective use of resources and having one body to speak and act for physicists, but a main one was to improve physics journals. *The Proceedings of the Physical Society* was a long-standing and respected journal but its scope was effectively limited to classical physics. In his view one of the responsibilities of the joint body would be to publish additional journals covering quantum, nuclear, solid-state, atomic physics, etc., as needed, and this indeed was done. His interests in scientific publishing were already well established. In 1948 he had become editor of the *Philosophical Magazine*, published since 1798 by Taylor & Francis, and his association with that firm lasted until the end of his life. He became chairman in 1970 and president in 1975 and greatly enjoyed his involvement with the firm. In 1959 he helped to found *Contemporary Physics (A Journal of Interpretation and Review)*. I was invited to be a member of the original editorial board (and am the only one still alive). Nevill formally joined the board in 1967, though from the beginning he gave his support and wise advice. When I retired from the board in October 1990, I was very touched by his kindness, at the age of 85, in travelling specially from Milton Keynes to the Taylor & Francis offices in London to attend the farewell lunch given for me by the firm. He contributed interesting and useful papers to the journal, ranging from the teaching of quantum phenomena to metal–insulator transitions and electrons in glass.

In November 1960 I introduced a session on specialisation and the curriculum at the first of the Gulbenkian Educational Discussions, with Sir Charles Morris, Vice-Chancellor of Leeds University, in the chair (reported in *Universities Quarterly*, Vol. 15, March 1961). I was concerned with the need for and provision of a range of first-degree courses in universities: special, joint and general honours, and pass. In the light of my own experience of teaching a variety of physics courses I outlined my views on how they should be planned and taught. Nevill was there and expressed himself as being in full agreement. This led to discussions between us over several years. He was very dissatisfied with Part I of the Natural Sciences Tripos at Cambridge, partly because it was an inadequate preparation for and left too little time for advanced work, and partly because the three subjects chosen by a student were studied quite separately. Largely due to his energy, persistence, determination and clear argument, changes were made after two or three years to enable future specialists to study some advanced physics (or chemistry) in their second year. But his interests in education, not only physics, were much wider than this, extending from school to postgraduate levels. I was greatly impressed by the thought, energy and time he devoted to educational problems. His contributions in this area can stand alongside his outstanding research in physics.

From 1977 I was Pro-Chancellor of Warwick University for fifteen years (and Chairman of Council for thirteen) and during this period met Nevill a number of times when he visited the Physics Department there. On one occasion, learning that he was to stay the night in the university, I invited him to have dinner with me in a university restaurant. I sent him detailed instructions on when and where to meet me. I waited in vain for him at the appointed place. With the help of the Physics Department I eventually found him at a different place, feeling somewhat put out because I was not where he had imagined I should be. Rudolf Peierls has related that Niels Bohr's secretary used to put travel instructions into each of his pockets so that he would find at least one of them, and perhaps this would have worked with Nevill. I remember General Pile saying, 'Professor Mott sometimes asked to see me and when he arrived he had usually forgotten what he wanted to see me about.' But we had an enjoyable dinner. Nevill praised the wine and its temperature, and we had an excellent discussion on a range of topics. One of them was the attitude of scientists to religion. We found ourselves largely in agreement, though he had thought more deeply and read more widely than I had. While not believing in the literal interpretation of physical miracles, for example, he now found comfort in the beauty of Anglican services based on the 1662 prayer book. He had thought deeply about the differences between scientific and religious truths and was able to subscribe to both. I took him to Radcliffe House in the university, where he was staying the night, and shall always remember him as he was that evening, relaxed, friendly, with a well-deserved sense of achievement, still mentally active, and above all fundamentally happy and content.

SIR ARTHUR VICK was born in 1911. He is a physicist – retired but still in love with the subject. In addition to the activities mentioned in the text, he has been a senior lecturer in Manchester University, the first Professor of Physics at Keele, a member of the University Grants Committee, and Vice-Chancellor of the Queen's University, Belfast.

Postwar Bristol

Memories from 1945 to 1954

Peggy Fisk

I was introduced to Professor Mott and his family in 1945 when his sister-in-law, Mary Horder, asked me whether I knew of anybody who would be able to move into Stuart House with her. Mary was working at that time and needed assistance to prepare the house for her sister and two small daughters to move in to live.

I was seeking accommodation at the time and readily agreed to move in, so Mary, my three-year-old son Tony and I moved into Stuart House, where we stained floorboards, painted walls and generally prepared the house ready for the arrival of the family. From an initial period of three months we eventually stayed for nine and a half years and when my husband returned from the RAF, he too was readily invited to join us. I often remember those years with great fondness. We were all very happy and after the war years the Royal Fort provided an ideal home for three young children, with lovely gardens and security whereby they could run around freely.

Sir Nevill was a very humble man, shy with a wicked sense of humour and an infectious booming laugh. He enjoyed taking the children on outings and my son always thought it a great adventure to go out with 'Uncle' at weekends. I remember particularly one such outing when they all went to Portishead and were allowed to wallow in the mud on the banks of the Severn Estuary. They returned home absolutely covered in mud from head to toe, much to Sir Nevill's amusement. All three children were immediately given a hot bath but emerged covered in red blotches from the effects of the mud – heaven knows what it contained! Another regular treat was to be taken over to the Physics Lab, where they were permitted to scribble and draw on the Big Blackboard, which they did with relish.

One day Tony, who by this time was eight years old, proudly announced, 'Uncle has taught me to drive!' I asked how this was possible. Tony told me that Sir Nevill had shown him the controls and what they did, then sat back in the passenger seat folded his arms and, with Tony having to stand up to see behind the steering wheel, promptly said, 'Right off we go!' Apparently Tony managed to drive Bluebird (this being the family name for the Morris 8 which was Sir Nevill's car at the time) around the oval-shaped driveway between Stuart House and the laboratory, which I hasten to add is private and free of traffic at the time. Sir Nevill was delighted and pleased with his role as driving instructor and thought the whole event enormous fun.

Sir Nevill would often assist with the washing-up and would be wiping a piece of china so vigorously I feared the pattern would be removed when he would suddenly walk off and disappear across to the laboratory. After a period of time, the house would become bereft of glasses, cups and plates together with tea towels and I would have to go over to his lab to retrieve them.

One day he arrived home wearing a new hat and, having asked everyone's opinion of the new purchase, enquired if we knew the whereabouts of the car. He was on the point of reporting it stolen, when someone asked, 'Where did you buy the hat?' 'Marsh's,' he replied and received the answer that the car could probably be found there. This was indeed the case and he was not amused to find a parking ticket on the windscreen as a result!

I recall a conversation nearly fifty years ago when he announced that mankind would one day walk on the surface of the Moon, adding perhaps not in my lifetime but certainly in that of our children. How prophetic those words have proven to be.

I feel privileged to have known Sir Nevill and his family, and I will always retain the fondest memories of a very rare combination of human values, love of the countryside which he shared with his family, and dedication to his work. I will always be grateful to him for giving me and my family such a happy home at Bristol.

MRS P. FISK was born in Northern Ireland in 1919 and arrived in Bristol in 1939 just a couple of months before war was declared. In 1945 she moved into Stuart House with the Mott family and kept in contact with Nevill for the rest of his life.

The Bristol Period: 1945 to 1954

Jack Mitchell

Beginning in 1933, with strong support from Professor A. M. Tyndall, Nevill Mott departed from traditional theoretical physics and became concerned with the processes of physical, chemical and photochemical changes in solids and their dependence on atomic properties and crystal defects. Following the work of Taylor, Orowan, Polanyi and J. M. Burgers in the period 1934–1940, he became interested in the role of dislocations in the plastic deformation of crystals. With Harry Jones, he applied band theory to the study of electronic properties of metals and alloys. With Ronald Gurney, he discussed the model for the F-centre in an alkali halide crystal and the mechanism for the photochemical decomposition of silver halide microcrystals which occurs during the exposure of photographic films. With Frank Nabarro, he discussed precipitation hardening in alloys. He also published two papers, *Theory of Formation of Protective Oxide Films on Metals* and *Reactions in Solids*. In all this work, one of his aims was to apply knowledge of solid-state physics and chemistry to both fundamental and industrial problems to stimulate relevant and effective research. He excelled in his ability to analyse the accumulated results of experimental studies of properties dependent on defects in crystals, to propose models for these defects, and to make physically based order-of-magnitude calculations for processes with the models which had observable consequences. These wide-ranging interdisciplinary activities were admitted as appropriate for a physics department at Bristol and this acceptance contributed to the rapid growth of solid-state physics there after 1945. In the atmosphere created by Professor Tyndall, good science was what mattered.

I came to know Nevill in 1944. I had become involved at the Armament Research Establishment at Fort Halstead, Kent, in the use of ultra-high resolution schlieren photography to study interactions between high intensity shock waves generated by explosive charges. The observations were analysed in collaboration with Kenneth Thornhill of his Theoretical Physics Division. I began attending a series of lectures which Nevill was giving on topics which he felt would provide profitable research areas when he returned to Bristol after the war. I was particularly interested in the lectures on plastic deformation of metals and alloys, on reactions in solids, and on the Gurney–Mott theory of latent image formation in silver halide crystals. I was enthusiastic about a number of projects which would match my own interests very closely. This led to my being offered a lectureship in experimental physics at Bristol by Professor Tyndall, and I began working there in September 1945.

After 1945 Nevill played a decisive international leadership role in the growth of the still emerging field of solid-state physics, through his lectures at home and abroad, his organization of summer schools and conferences, and his interactions with industrial research laboratories. These activities brought a steady stream of frontier research workers to Bristol, who gave stimulating lectures which were followed by vigorous discussions. They also attracted graduate students and postdoctoral research associates from other universities in the United Kingdom, and from overseas countries, including René Bourion, Jacques Friedel, Dick Polder, Doris Kuhlmann (later Kuhlmann-Wilsdorf), Alfred Seeger, Jim Allen, Jock Mackenzie, John Rivière, Huang Kun, Peter Schroeder and Jan Van der Merwe. This resulted in the rapid growth of the different research groups to an overall operation of more than the critical size needed for effective creative activity. There were many summer schools or conferences in Bristol in the period 1947 to 1954, on the strength of solids, properties of ionic solids, crystal growth, fundamental mechanisms of photographic sensitivity, semiconductors and transistors, and defects in crystalline solids.

Nevill and Ruth Mott moved into Stuart House opposite the main entrance to the H. H. Wills Physical Laboratory. They were always warm and welcoming hosts, and on many occasions small groups retreated from the laboratory to the drawing-room for discussions of problems and projects, often over a glass of sherry. This proximity to the laboratory and the many social activities involving students, faculty, and visitors helped to establish the relaxed atmosphere, conducive to new discovery, which characterized the Bristol period.

The tearoom on the third floor of the laboratory was an essential element in the maintenance of this vitality. Many new ideas were proposed and elaborated on the blackboard at the 4 P.M. sessions. After January 1947 Charles Frank was a regular contributor. His dynamic galloping dislocation model for the formation of slip bands was presented here and critically discussed by Frank Nabarro, who was then a Royal Society Warren Research Fellow. During the course of a discussion on the role of kinked surface terraces on sodium chloride crystals in crystal growth with Keith Burton and Nicolas Cabrera, I asked Charles if he could account for the formation of very thin needles of sodium chloride which I had seen above the high-water line in cavities in basalt around Lyttleton Harbour. Uncharacteristically, there was no immediate response. That evening, while browsing through a book on early Mideastern architecture, he came across an illustration of a ziggurat tower with an outer upward-climbing spiral terrace, and there was born the screw dislocation theory for the growth of needles and of more general crystal growth.

The Christmas party in the main lecture theatre was always a memorable occasion with Professor Tyndall dressed as Father Christmas. One year he arrived seated inside a large Faraday cage with surrounding flashing sparks. Another year, he gave a copy of the poems of Robert Burns to René Bourion, who had succeeded in igniting a hydrogen jet with a cigarette and seriously

Postwar Bristol

burned his arm. But most memorable was his conferring the Order of Magnitude on Nevill Mott which brought the house down. The Order was a roughly torn large oval of brown card with an order for vegetables which he solemnly hung around Nevill's neck. These were happy days of *resorgimento* during which we found time to relax during some afternoons and to go for long walks at weekends while working far into the night.

In 1945 after taking available facilities into account, we decided to initiate a programme of experimental research on the initial stages of oxidation of aluminium and copper, closely parallel and directly relevant to a theoretical programme on the formation of oxide films on these and other metals. Nevill had proposed that oxygen molecules, adsorbed on a clean metallic surface, or on the surface of a thin oxide film, are dissociated and charged negatively by electrons from the metal. Cations, or vacant cation lattice sites, drift across the film in the field thus established, so that cations can combine with the oxygen ions at the surface to build an oxide film up to a limiting thickness determined essentially by the electron tunnelling distance. Observations of the adsorption of oxygen on clean surfaces were needed for the evaluation and extension of this theory.

Because of the need for initially clean surfaces, the experimental systems had to be produced by the condensation of thin films on substrates under ultra-high vacuum conditions. For the work on adsorption with thin films of copper, aluminium and other metals, a self-gettering system was introduced by Jim Allen in 1948 with a spherical bulb and a thoroughly outgassed heated central bead as the source of the atoms. This system was subsequently used for the study of the properties of thin films of metals by Glynn Holloway, Eric Dorling and Chris Evans. The measurements of adsorption and of initial rates of low temperature oxidation were consistent with the theory for aluminium, and confirmed the electron tunnelling mechanism for the formation of a protective oxide film with growth up to a limiting thickness. The results with copper at low temperatures were not consistent with the theory and this stimulated further theoretical work by Nevill, who published the third part of his series on the formation of protective oxide films on metals, and by Nicolas Cabrera, who came to Bristol from Paris in 1947. Their review article *Theory of the Oxidation of Metals* was published in 1949.

In parallel with this work on oxygen adsorption and low-temperature oxidation of thin films of metals, methods were developed for measuring work functions and the change in work function accompanying the adsorption of oxygen on the surfaces. René Bourion improved the unsaturated diode method. A major advance was made by Bill Mitchell in 1948. He introduced a focusing electron gun method for the measurement of the work functions of copper, silver, aluminium and germanium. The Kelvin method was used by John Rivière in 1952 to measure contact differences of potential between thin films of copper, silver, aluminium and gold, taken in pairs, and between silver or gold and nickel, tungsten, molybdenum and iron. He obtained self-consistent results with a high level of reproducibility. Peter Schroeder measured the work functions of copper and silver by the photoelectric method, obtaining values in excellent agreement with those of Bill Mitchell. With this method it was possible to obtain reliable values for the contact differences of potential between clean surfaces and oxygen-covered surfaces, one of the objectives of the programme. These studies on the properties of thin films of metals were made possible by the outstandingly skilful glass-blowing and good-natured perseverance of John Burrow, BSc.

Early in 1948 I was encouraged by Nevill to take a serious experimental interest in silver halide systems. It was evident that new techniques were needed. Thin sheet crystals of silver bromide were made by crystallizing molten discs between glass plates. The exposure of crystals of high purity and perfection did not produce either a surface or internal latent image. The crystals had to be sensitized with silver oxide, or with silver oxide and silver molecules, before exposure produced a latent image that could be developed with either a surface or an internal developer. John Hedges, Trevor Evans and Peter Clark showed that the thin sheet crystals could be sensitized by all the methods of silver halide photographic technology. Douglas Keith studied the processes of development. These crystals provided a model which reliably reproduced the properties of the silver halide microcrystals of photographic emulsions. The experimental work suggested that the latent image was formed initially, not by the primary photolysis of the silver halide with release of silver and halogen atoms, as in the Gurney–Mott theory, but by the photoaggregation of silver atoms, chemically equivalent to the molecules of chemical sensitization. Nevill followed this work with keen interest and there were many stimulating discussions of its implications.

The most important observations with thin sheet crystals were first made in November 1952 by Hedges and Mitchell. Lightly annealed crystals of silver chloride or silver bromide, sensitized with silver oxide and silver molecules, were darkened by exposure. When they were examined with a high-power microscope, it was seen that what had appeared to be high-quality single crystals had a substructure of polyhedral grains with planar interfaces and dimensions of the order of microns. On the subgrain boundaries, continuous three-dimensional networks of dislocation segments, intersecting in triple nodes, had been made visible by decoration with closely spaced particles of photolytic silver. It was clear that the precise nature of the mosaic structure of a real crystal, which had been envisaged by Darwin as early as 1914, as an assembly of slightly misaligned blocks, was being observed for the first time. As soon as we saw the detailed dislocation structure of the subboundaries, I asked John Hedges to see if he could find Nevill. I shall never forget his delight and enthusiasm when he first scanned the dislocation networks in these crystals. With the very limited depth of field of a large aperture high-power microscope objective, it was possible to scan the arrays of dislocations by focusing the microscope slowly up and down. This gave a vivid three-dimensional impression of the dislocation structure of the subboundaries, which it was unfortunately not possible to reproduce in two-dimensional photomicrographs. The adjacent subgrains were rotated through very small angles relative to any particular subgrain. Certain relative orientations gave small-angle tilt boundaries formed by parallel arrays of equally spaced edge dislocations. Others gave small-angle twist boundaries with hexagonal arrays of dislocations. When the crystals were plastically strained then exposed, high-density arrays of glissile dislocations were seen on glide planes within the polyhedral subgrains, leaving little doubt that the unit displacement processes of plastic deformation were being observed in operation within crystals for the first time. With these observations, the widespread doubts about the physical reality of dislocations vanished overnight and the theory of dislocations surged forward.

The theory of dislocations had been steadily refined and extended after 1945 when Frank Nabarro returned to Bristol for four years to resume his work with Nevill on properties of dislocations as linear lattice defects with elastic properties, solute and precipitation hardening, and transient creep. The Frank–Read dislocation source was proposed in 1950 but a Frank–Read source was never observed in a silver halide crystal. This source was introduced by Nevill in theoretical mechanisms for fine slip, slip band formation, and work hardening in metals. The basic concepts of dislocation theory were clearly presented by Charles Frank in 1951 and Norman Thompson introduced the notation of the Thompson tetrahedron and discussed the nodal structure of dislocation intersections in face-centred cubic crystals in 1953. With this notation, Frank analysed the detailed structure of the hexagonal arrays of dislocations observed by Hedges and Mitchell.

The postwar Bristol period came to a close in July 1954 with the conference on defects in crystalline solids, which I organized as a tribute to Nevill. He had created an atmosphere in which research could flourish and he had stimulated strong interactions between the research groups with warm and appreciative understanding of the significance of their achievements. Beyond the academic activities, his good relations with industry helped to provide financial support upon which the research programmes depended.

PROFESSOR J. W. MITCHELL, FRS, was born in Christchurch, New Zealand, in 1913. He was Reader in Experimental Physics at Bristol from 1949 to 1959. At the University of Virginia, USA, he was William Barton Rogers Professor of Physics (1965–1979) and since then has been a Senior Research Fellow.

Recollections

Charles Frank

I knew Nevill Mott during the war, but not very much. I knew about his operational research work for Anti-Aircraft Command. In particular, I was familiar with some of his work on fragmentation of shells, foreshadowing his interest in deformation and fracture of metals, which he was to develop after the war, in contrast with his highly abstract theoretical physics before that – and his solution to the problem of why coastal anti-aircraft batteries scored so much better in 'rounds per bird' than those entirely surrounded by land. He discovered that their advantage disappeared once you counted those shot down only on the land, those reported as falling into the sea being so much less reliably confirmed.

Throughout the war I had been telling Eric Rideal in Cambridge that as soon as this little thing was over I would return to his lab. Come the end of the war he confirmed that I should come back to his lab, but it would not be in Cambridge any more, it was moving to the Royal Institution. As I had just had six years in London, not the best six years for enjoying its advantages, an invitation from Nevill Mott to come and have a look at Bristol seemed attractive to me. Here he took me to the top of the laboratory tower to look down on the campus, and said, 'This is a university of 1600 students; we plan to let it grow to 2000 and then stop. We think that would be a nice size for a university.' It must have been most of ten years later that John Shepherdson, one of the mathematics professors, and I wrote a letter to the Vice-Chancellor to tell him that the university must be allowed to grow to 8000 students so as to maintain the quality of its library.

I did not perceive the plan at the time, but what Mott was doing was recruiting people for a department of theoretical physics with practical and possibly industrial applications for what it did. I think he had already recruited Eshelby and Nabarro. His technique in guiding us into particular subjects was to give a lecture course presenting the state of the art in that subject and then leave it to us to see how it could be developed. Thus he gave us the theory of plastic deformation as it had been left by G. I. Taylor, i.e. with dislocations as the moving elements of deformation, and mosaic boundaries as the obstacles that impeded them. It was for me to say that the mosaic boundaries could merely be arrays of dislocations, and so forth. Between us, Eshelby, Frank and Nabarro made a strong team for exploring this theory of plastic deformation. Nicholas Cabrera had already done some joint work with Mott on surface oxidation of metals. Keith Burton had been sent by ICI to develop a theory of crystal growth: one of the rare cases of a directed research producing its pay-off exactly on target.

On the various topics we explored between us, we arranged annual courses which came to be known as Mott Schools and which began with a number of lectures on the newly developed doctrines mostly from the Bristolians but with some contributors from elsewhere. It was as a contributor from elsewhere that Alan Cotterell first became one of my acquaintances. After these lectures there would be one or two presentations of papers giving examples within that subject area.

Mott left Bristol to go to the Cavendish Chair in Cambridge in 1954. I remember him coming to our then newly acquired cottage with a rather glum face to tell us of his impending departure. He said he was virtually conceived in the Cavendish and if his father and mother had not still been alive he might have been willing to refuse the offer of that chair and stay in Bristol, but they would never understand him doing that, so he had to go.

The Motts lived in Stuart House across the lawn from the Physics Department, thus 'living over the shop', as he once put it. He was frequently able to organise social gatherings at short notice. He would call us in because he recognised that my wife had the ability to make a sticky party go, and he would also occasionally take us in his little car out for a walk on the Mendips or the Cotswolds above Wotton-under-Edge. But I think I will leave it to my wife to finish off this memoir with more recollections of Nevill; in particular, my wife became a partner in toy making with his sister-in-law, Mary Horder, who married Jacques Friedel, subsequently Président de l'Académie des Sciences in Paris.

Postscript by Maita Frank

I met Professor Mott in autumn 1946 when we came to Bristol for Charles to see the Department of Physics with a view to taking a job. Towards the end of our visit, Nevill said that he would invite us home for the evening but he and his wife Ruth were invited to a large party at Esme Bain's he would not wish to miss. But after a moment of thought he proposed to 'gatecrash' us, as his wife decided not to go and he was sure Esme would not mind. Also that it would be a good hunting ground for us to find somewhere to live. He was so right! We found a flat for six months and made friends with three Bristol families for life.

SIR CHARLES FRANK, OBE, FRS, was one of Nevill Mott's first batch of appointments after the Second World War to the Department of Physics at Bristol. He subsequently succeeded (with Maurice Pryce and Cecil Powell intervening) to Head of Department from 1969 to 1976.

Some Personal Reminiscences

Bill Mitchell

Or as Shockley said when visiting Bristol, Little Mitchell, thus distinguishing the youthful athletic figure from his sturdier but well-proportioned supervisor – more recent acquaintances please note!

First Encounter

I first met Nevill Mott in 1947 when I was a research physicist in the research laboratories of Metropolitan-Vickers in Manchester, having gone there in 1946 after a wartime radio degree at Sheffield. That meeting was brief, perhaps twenty minutes, as Mott made his rounds of the semiconductor laboratory guided by Dr R. W. Sillars. Sillars' part of the laboratory had made its name developing Metrosil, a non-linear resistor for spark quenching and for lightning arrestors. It was fabricated usually in disc form from silicon carbide grains held in contact by a ceramic matrix. Sillars had asked me to measure the $I–V$ curves of single-crystal contacts between similar crystals, in fact a p/p contact. I was immensely excited at having a real research task.

Sillars and Doidge had developed a high-impedance electrometer bridge for measurements of the internal resistance of individual crystals and I was able to adapt this for contacts between crystals. The results all showed non-linearity and a small rectification due, I claimed, to different levels of p-character in the two crystals. Mott chatted extensively about this and about my view that surface states were enhancing the space-charge barriers, thereby giving slightly asymmetric mini-Schottky barriers whose theory Mott had given. He explained that his collaborator, Miss Dilworth, was working on the theory of contacts between semiconductors and that they would let us know what they did. This encounter early in 1947 led to Sillars and Mott deciding that Mitchell should go to Bristol. I arrived in January 1948, joining J. W. Mitchell's group on the surface properties of metals, charged first with measuring the work function of clean copper.

Bristol

Bristol in 1948 was a revelation to me. In the previous year, Powell and his fourth-floor colleagues had discovered the π-meson; in 1948 Fröhlich and

Nabarro were still around, before moving on to Liverpool and Birmingham respectively. Fröhlich at colloquia, including those of research students, seemed aggressive but was actually quite nice and Nab, conversely, seemed quite nice but was actually quite tough. Mott's postgraduate lectures were research seminars in themselves, including the localisation and electron density graph which appeared in PPS around 1949, inspired by the properties of NiO and Ni, and which was to come to the fore twenty-five years later. Frank and Cabrera were formulating their dislocation theories of crystal growth and J. W. Mitchell was making good progress with his theory of the photographic process. This was really closer to JWM's heart than the surface of copper, but he never allowed it to show.

Mott encouraged these interests, so the department was infused not only with Nevill himself, but also the people he had gathered around him.

Pauling came during one of my three years (1948–50) and there was a huge audience in the large Chemistry Department lecture theatre. He was billed as talking about the electron theory of metals, and he and Mott were expected to cross swords. Disappointingly they agreed early on that they were looking at metals from different points of view and that neither was right or wrong. Nevertheless, Pauling found many opportunities to manipulate his six-inch slide rule, taken from his top pocket.

Nobel Prizes *et al.*

Mott proceeded through solid-state physics: scattering theory, wartime work on alloys and Duralumin, metals and alloys, semiconductors and insulators, amorphous semiconductors, and finally high-temperature superconductivity. Few people can have had such seminal influences on so many major fields, backed up by his chosen collaborators: Massey, Jones, Gurney, Davis, and so on. This corpus is nowhere rivalled and it is a mystery that it only produced one Nobel prize!

Indeed, a few years later, at a *very* relaxed lunch in New College, Nevill ruminated on how difficult it would be to be awarded two Nobel prizes in the same subject. Oddly, Willis Lamb on a summer visit to Oxford, at dinner, had also spoken in those terms. At Mott's lunch we all came up with Bardeen. Then what about Madame Curie? No, one chemistry and one physics. And Pauling? One was for peace!

In September 1990 Mott gave a brilliant evening lecture at the high-temperature superconductor conference in Cambridge. He was probing uncertain aspects of his approach to the copper oxide superconductors, stimulating people to take up these problems, just as in 1947 his postgraduate lectures had dealt with the Cu_2O – metal rectifier. I had the feeling that if Mott was involved with the oxides of copper, which still had not given up all their secrets after nearly fifty years, all was right with the world.

The Kronig–Penney Model and a Linear Journey

Another of the reference points to the research students of Bristol was the Kronig–Penney model – Mott and Jones stuff and a postgraduate lecture topic. Some of us had heard of Penney because of his wartime exploits and his role in the production of the UK atomic bomb in 1946. Indeed, as his career became known, its great characteristic was the reduction of complex problems (e.g. blast measurements from atomic explosions) to those having tractable and simpler solutions. Certainly the Kronig–Penney model has this characteristic.

In 1964–65 I found myself with Mott in Groningen, walking along streets without pavements on our way to have dinner with Kronig, who was President of the University, if my memory serves. I was still blessed with a youthful heart, excited that I was going to meet another of the pillars of my early solid-state physics, and in the company of the man who had introduced me to band theory. Nevill had been appointed by Roy Jenkins (Secretary of State for Air) to conduct an inquiry on the relationship between the Ministry's research establishments and the universities. Nevill had asked Eric Eastwood and myself to join him, and as well as visiting Farnborough and Malvern, we visited Philips and universities in Holland, because the Dutch seemed to have established very effective relationships. It was just Mott and me in Groningen, and we decided to walk to Kronig's office. As contributors to and readers of this book will know, Nevill could become intellectually totally immersed in the physics problem currently on his mind, totally oblivious of his surroundings or other matters. On this occasion the resistivity of liquid mercury took over. 'Well, can't you measure tau near the density of states minimum?' As these questions came, I was trying to keep within earshot by jumping in and out of the traffic to avoid being run over on our unprotected pedestrian strip. Mott seemed totally unaware of the predicament he had placed me in, made worse by the difficulty of answering the question. Eventually we arrived and were shown to Kronig's office. After some general conversation about what we were doing, Mott said we had been discussing the resistivity of mercury on the way over, but the traffic had been terrible, 'Bill was nearly knocked over several times.' So he hadn't been totally immersed after all. Advocating joint appointments and research collaboration, we produced a rather good report, which preceded a similar document from the Royal Society.

Finale

I used to meet Nevill from time to time and would always find it stimulating; he would always ask about and listen to what you were doing before finding a way of linking that to some aspect of his own interest. He was respected throughout the world and his influence on UK solid-state physics was colossal.

SIR WILLIAM MITCHELL, CBE, DSc, FRS, took a wartime degree at Sheffield in physics and radio, before following a research career at Metropoliton-Vickers in Manchester, then at the universities of Bristol and Reading, before being appointed Professor of Experimental Philosophy and Head of the Clarendon Laboratory, Oxford, in 1978. He was Chairman of SERC from 1985 to 1990 and President of CERN from 1990 to 1992. He is currently Emeritus Fellow of Wadham College, Oxford. His research interests included the use of neutron scattering to investigate defects in crystals and the local order in melts.

Postwar Days at Bristol

David Gibbs

Shortly after the beginning of the war, after finishing off some work on neutrons and nuclear fission with a group under G. P. Thomson, I worked on the development of naval radar transmitters. At the war's end, I was eager to get back into academic life (in spite of its great financial disadvantage) and was fortunate to be given a lectureship in the Physics Department at Bristol University. I suppose I had become more of an electrical engineer than a physicist, so I was eager not only to exploit electronics in the cause of physics, but to repair my rather shaky foundations of theoretical physics.

I was soon to encounter Professor Mott, as he then was, in several different ways. I recall him as one of the kindest people I have known, though I suspect that he had some doubts about my suitability as a member of the department. His response to any remark or question was always a smile, and it was never a sign of condescension, rather, I felt, he was giving himself a moment to compose the clearest and most down-to-earth reply. It was this attitude which made technical discussions so fruitful, the basis of the problem was always made clear at the beginning, in elementary terms, and development was without impatience or interruption.

For a theorist of such profundity, he was very practical. In the middle of one of his talks, I recall his saying something like 'Does anyone happen to know the tensile strength of nickel?' Someone produced an enormous number of dynes per square centimetre. 'Oh,' he answered, 'and what's that in proper units, tons per square inch?'

There was at the time a flourishing Students' Physical Society, and I ventured to give them a talk on semiconductor rectifiers. This was just before the transistor era, and there were no longer many crystal sets about, but silicon crystals were universally used in radar receivers, and for years we had used 'metal rectifiers' to provide high tension for our valve wireless sets. I knew quite a lot about the electrical characteristics and application of these devices, but was ignorant about the then rapidly developing theory of their action. I was touched, and secretly much relieved, when Mott apologised to me that he was unable to come to my talk.

There were, of course, many lectures and seminars, but there were also less formal discussions in the tearoom, and at his house opposite the laboratory, where there were often distinguished visitors such as Heisenberg. He and Mott discussed old times, and recalled that Dirac had studied electrical engineering in Bristol. 'Oh yes,' said Heisenberg, 'I was most impressed when this clever young man came and sat at my feet in Berlin.' 'Come off it,' said Mott, 'I seem to remember from the Cambridge days that all three of us were about the same age – he couldn't have been much younger than you.' 'Oh yes, he *was*,' said Heisenberg (you must imagine his slight German accent). 'He was at *least a year* younger!'

There was no Professor of Music in the university then, but considerable musical activity, by both staff and students, which the Professor of Greek, H. D. F. Kitto, and his wife did much to foster. There was a highly erratic but enthusiastic orchestra, a good many of whose members came from the Physics Department, and once when deputising as its (semi) conductor, I was surprised to find Mott playing the clarinet. Unfortunately, I think he became too busy for this to continue for very long, but other members of his family contributed greatly to the musical scene.

Although office parties seem to have a bad reputation, I am sure that a good Christmas party can work wonders for the spirit and unity of any organisation. August figures can make fools of themselves, and the humbler can display quite unsuspected talents. I recall these parties in the Royal Fort with the greatest pleasure, and I remember that at one of them Mott gave a travesty of a lecture, with the head of the department acting as his assistant. I helped prepare a set of fake demonstrations, each of which produced the reverse of the effect which physics would lead us to expect. Perhaps it was not so very different from the real thing!

DAVID F. GIBBS was born in Gillingham, Kent. His schoolboy obsession with such things as wireless sets and fireworks predicted a career in practical science. He graduated in physics at the Royal College of Science in 1938. From the end of the war to the present he has been at the Physics Department of Bristol University, apart from a year spent at the Bell Telephone Laboratories in New Jersey. He was Senior Lecturer when he retired.

Master, Colleague and Friend of our Laboratory*

Yvette Cauchois

On 30 September 1995, Sir Nevill Mott was 90. Among the messages and celebrations, he had received the good wishes of our laboratory. He was kind enough to answer; in his reply, addressed to Professors C. Bonnelle, A. Maquet and myself – who had directed the Laboratory of Physical Chemistry in Paris after its founder Jean Perrin – he wrote, 'It has been a great pleasure to know your group over the last half century.' Indeed in his autobiography, *A Life in Science*, Sir Nevill Mott recalls that he was able to visit Paris again in 1947 after the Second World War.

Do allow somebody surviving from that already far gone period to tell a little more about Nevill's first contact with our group. The reputation of the British visitor, then Professor at Bristol and elected to the Fellowship of the Royal Society, was already well established. I do not recall why or how the CNRS, the national science organisation in France, was led to provide him with a car and a driver. I do not remember if I was involved. The fact is that I had the privilege to accompany him to Chartres in that car. I still feel my joy on leaving town, in seeing trees again along a nearly empty road, green fields where a fallacious technique had not yet prevented cornflowers and poppies from flowering, in seeing the cathedral grow on the horizon, a cathedral miraculously saved from bombs aimed at the nearby airport. Our visitor appeared to have some idea about this miracle. He had probably chosen our destination himself; he was particularly fond of Chartres, just as he also liked the remains of Romanesque or Cistercian art saved from the criminal destruction of wars and revolutions. His adolescent wanderings had left him with many memories, and a stay in Lausanne to learn the rudiments of French had given him a taste for our language. He enjoyed speaking in French; in our further correspondence, I wrote in French and he in English with some snatches of our language.

To return to our visit, we found ourselves in front of the cathedral whose old stones were still alive but which was lacking its marvellous glass windows. The empty apertures were protected by long white hangings which flapped with the wind coming from the Beauce. The next year, I believe, most of the glass windows were back in place. Nevill was to visit the sanctuary again on the occasion of further travels.

In 1947, on the return journey, despite my shyness, I was able to talk shop with him, encouraged by his smiling kindness. This was, for our labor-

* Translated by J. Friedel.

atory, the beginning of a lasting and fruitful relationship with this remarkable personality. Our research work interested him. We published together in the *Philosophical Magazine* in 1949, a paper bringing together X-ray spectroscopy and solid-state physics.

In his book, Nevill says that the mixture of physics and chemistry at the Laboratory of Physical Chemistry in Paris, intended by Jean Perrin, looked to him favourable for developing solid-state physics, but this could not have developed normally during previous years. He helped us along those lines by accepting to give, in our lecture room in 1950, a much valued series of talks. His lectures were collected by us into a booklet which was for some time much sought after. In this way, he became our Master.

In 1951 he wrote a preface to my small book *Les Rayons et la Structure Electronique de la Matière*. In the chapter on electronics of solids, analysed by X-ray spectrography, I had referred to the results by M. W. B. Skinner on long-wave X-ray bands, as analysed theoretically by Mott (1936). The book by Mott and Jones was always on my desk. Among our research people, the volume by Mott and Massey in its successive editions, was especially appreciated. We also possessed the book by Mott and Sneddon of 1948, especially concerned with applications more than with the basis of wave mechanics.

However, it would be Jacques Friedel, when becoming professor in Orsay, who after spending three years in Bristol with Nevill, succeeded in creating in Orsay a centre of solid-state physics which acquired a worldwide reputation.

Nevill Mott's teaching activities went further in connection with our laboratory. He had decided to rewrite his first book, *An Outline of Wave Mechanics*. Published in 1930 it had been very successful but was now out of print. For our students, at a time when they were not accustomed to reading specialised English books, I had the pleasure of preparing a French version of this new book. With his customary generosity, Nevill sent me his manuscripts as soon as they were written, so that the French and English editions could appear simultaneously (this wish was realised, if one takes into account the respective delays due to the editors).

The French edition appeared in 1953, under the title *Eléments de Mécanique Ondulatoire**. It was much studied by our students. For me, the times of isolation when I worked with this text, on the bench of a park or sitting on the grass in the country, are some of the happiest moments of my professional life. At a higher level the book by Mott and Sneddon was very useful to me.

In 1954 the Science Faculty of the University of Paris offered Nevill Mott a doctorate *honoris causa*.

Our laboratory was pursuing its activities in various fields, with increased personnel and material support. With the initial help of Manne Siegbahn, we

---

* The name *quantum mechanics* was not yet much used. The theory of the wave-like nature of the electron, founded by Louis de Broglie, was in everyone's mind.

had extended our studies of X-ray spectroscopy to the far-ultraviolet range. Research groups were investigating possible photographic emulsion required for our spectrographs and for detecting traces of ionising particles. The creation of the 'latent' image had been studied by Mott; and the book by Mott and Gurney was heavily used. While in the States, I went to visit the Kodak Center at Rochester; this preventing me from visiting the Niagara Falls, a fact for which I have now a lasting regret!

We were also able to introduce in Europe the use of a new source of electromagnetic radiation, synchrotron radiation; first by using the Italian synchrotron in Frascati, then with the storage rings of Orsay.

During all his years in Bristol, Professor Mott, who was knighted in 1962, always followed our efforts with unfailing enthusiasm. In 1954, when he left Bristol to become Cavendish Professor of Physics in Cambridge, then Master of Gonville and Caius, it seemed to me that his tasks became overwhelming; but he never neglected us.

Leaving aside the local problems discussed in his book, I would like to stress his constant preoccupation with the control of nuclear armaments. In 1940 physicists knew that nuclear fusion, with its production of neutrons able to start a chain reaction, could potentially be exploited militarily. The impact of this was explored in astrophysics but it seemed that great technical efforts would be required to switch from the idea to the actual fabrication of a bomb. However, in the United States, the Manhattan Project was launched in secrecy and with no lack of funds. Sir Nevill Mott had not taken part in the Manhattan Project. He was the leading spirit and president of an association of nuclear scientists whose aim was to look for realistic proposals for the effective control of nuclear energy. This role is described in his book. One also learns that he and my friend, the crystallographer Kathleen Londsdale, were the only people opposing access to Great Britain of nuclear arms. I just happened to be staying with Mrs Londsdale in the suburbs of London, when waiting for a train to go to her laboratory. I learnt about the bombing of Hiroshima. The illusion of a science beneficial to humanity and striving to serve only pure knowledge sank into horror. Sir Nevill took part in the Pugwash gatherings and organised one of them in Cambridge. These meetings were aiming to bring together scientists from the USSR and from the West, to lead to the control of nuclear armaments.

After voluntarily leaving the Mastership of Gonville and Caius, he could work again on his research. A group from his laboratory was able to show the motion of dislocations under the electron microscope. Dislocations had been studied in Bristol, especially by Charles Frank. Such research has brought together our laboratory with that of W. G. Burges in Delft.

When Mott acquired more free time in Cambridge, his knowledge and his perspicacity led him to develop the study of the properties of non-crystalline semiconductors. These are amorphous solids with innumerable applications. His well-known work in this field led to his receiving the Nobel prize in physics. Our colleague Christine Sénémaud, Director of Research at the

Centre National de la Recherche Scientifique (CNRS), studies amorphous silicon in our laboratory and had the chance to meet Sir Nevill Mott at a conference on amorphous solids; she is the last of us to have been able to see him again. After the discovery of superconductors at high temperatures, Sir Nevill proposed a theory for them. Until his last days, he remained true to his dedication for research and retained all the vigour of his thoughts. He also devoted himself to theology.

From the deep isolation that illness imposes on me, I have tried to write a last homage to the friend, now silent, whose silence we listen to. Beyond the dangerous academic formalism, my wish is that the young generation, which will hold the future of our laboratory, remembers Nevill Mott as one of its masters, a master whose wisdom can still lead you to the threshold of your own mind.

YVETTE CAUCHOIS is Honorary Professor at the Université Pierre et Marie Curie and Honorary Director of the Laboratoire de Chimie Physique de Paris. Her research interests are in X-ray spectroscopy of gases and solids, the study and use of synchrotron radiation, the optics of crystals and the use of bent crystals for obtaining X- and X-ray spectra (Chauchor's spectrometers). She is also interested in soft X-ray imaging of the sun (from rockets), space radiation and the history of science and philosophy.

Nevill in Bristol During the Early Fifties

Jacques Friedel

Introduced by my cousin, Charles Crussard, I first met Nevill in 1949 in the Metallurgy Department of the Paris School of Mines. Nevill was one of the first and foremost British scientists to roam around the Continent in the years after the war to establish contacts in research. This he did all his life, visiting many research centres and not only the universities which gave him honorary degrees! In France he made frequent visits to Y. Cauchois for electrons in metals from X-ray techniques, then to P. Hagenmuller and his group in Bordeaux, Mrs Schlenker in Grenoble, G. Amsel in Paris for their various works on oxides. He came, of course, a number of times to Orsay and we were planning to receive him in the autumn of 1996 with Alexandrov.

At Bristol, in the Royal Fort, from 1949 to 1952, I found one of the brightest centres of research and I arrived in time for a famous meeting on dislocations and crystal growth, which Nevill introduced with his usual force and simplicity. Coming from a rather depressed Continent, it was exhilarating

Jacques and Mary Friedel with Nevill and Ruth in Friedel's flat, Paris (1989).

to live in a place where Charles Frank had a new idea every week and Nevill himself switched his interest from electrons in metals to dislocations or to defects in ionic solids, while the fourth floor of the Royal Fort was preparing for Powell's Nobel prize in high-energy physics. A constant stream of visitors and new PhD students gave a strong international flavour to this Mecca of solid-state physics, established with Nevill's help well before the last war. New faces and old acquaintances constantly appeared for afternoon tea at the fourth floor, or above the Music Department for the morning coffee of the theoreticians. Ruth and Nevill frequently invited visiting guests to coffee parties in Stuart House, their charming eighteenth-century residence in the grounds of the Royal Fort.

Nevill tried at first to shift me to Charles Frank, who was then doing wonders on dislocations. I had come to learn about electrons in metals and I insisted on working with Nevill himself. For me, therefore, Mott was first of all the kind but critical scientist who taught me all I know of modern physics, by his books, his lectures, his discussions. He led my first steps in theoretical research by asking me to look in more detail at the electronic structure of metallic alloys, a field he had opened before the war, apparently a simple application of the scattering theories developed at length in his book with Massey. This proved a rather hard nut to crack before it finally developed in many directions. Always ready to hear of a new and interesting result, Nevill had not much patience with fumblers or fools; and it was at first a somewhat intimidating though stimulating experience to come into his office and talk about my research. Seeing my initial difficulties, Nevill suggested I should find another research subject, saying that solid-state physics was an exhausted field: that was before many future Nobel prizes including his own! He was harping on the future of physics in biology, a field which he was eventually to encourage his grandson, Edmund, to choose for his PhD.

When the time came for me to write my first theoretical paper, Nevill took great care in its production and made me rewrite again and again what was at first, I must admit, a rather ugly duckling! It is with him that I learnt the importance of a paper's presentation for its impact. My work was duly published in the *Philosophical Magazine*, the journal which Nevill had saved from oblivion some years before and which is still prospering with Taylor & Francis. I was then asked fairly frequently to act as a referee for proposed papers; my experience with Nevill in this field helped me later to launch for France the *International Journal of Physics and Chemistry of Solids*, then to renew our *Journal de Physique*. It was also during my stay in Bristol that *Advances in Physics*, the review journal, was launched. The occasion was a paper sent by his American friend Fred Seitz on the work of his group; this was too long to be published in the *Philosophical Magazine*. The success of *Advances in Physics* showed that it corresponded to a real need of the time.

A theoretician who expressed original ideas in simple mathematics, Nevill loved to talk to experimentalists and most of his research had practical importance. Already in the early 1950s, this research was on tremendously varied themes, from atomic and particle physics to electrons in metals, their transport and magnetic properties, from the photographic reaction and semiconductors to plasticity of metals and alloys.

Mott's strength was indeed to concentrate his mind on some recurring questions, which he attacked again and again on various sides. In the late 1940s he was interested in the way electronic repulsions could lead to a crystallisation of metallic electrons. This Verwey–Mott transition between a metal and an insulator would later become part of the reason for his Nobel prize. In his book with H. Jones, Nevill had been one of the first to point out the importance of the correlations of motion of metallic electrons which make them form a liquid more than a gas. He had given in 1949 a rough criterion for the crystallisation of that liquid, based on his prewar work on alloys; he was anxious to improve on it by incorporating some of the changes I had brought to the picture. I remember a car journey to Oxford where Nevill, while driving in his inimitable if somewhat hazardous way, constantly worried about this theme. On the way out, I was full of excitement about what new criterion should be obtained, what would happen if, as I thought then possible, the electron captured in its own potential well, was magnetic, etc. On the way back, while Nevill went on as a dog with an old bone, I was exhausted and barely recognised the pretty hamlet of Faringdon where we had stopped for lunch on the way out.

During this Oxford visit, where Nevill gave a talk, I met many new faces, in particular M. Price, the future successor of Nevill in Bristol, and my friend A. Abragam. What struck me most was a talk with Hume-Rothery. Stone deaf, he was asking Nevill all sorts of questions about electrons in metals and alloys and provided in return many experimental facts neatly put together according to his famous factors of atomic size, valency and electronegativity. I am certain this connection, lasting since well before the war, was essential to Nevill's grasp of metallurgical problems.

This connection with metallurgy, strong as it was at the time, was far from being Nevill's only interest in practical consequences of his work. He thought, quite rightly, that the industrial applications should soon replace, for solid-state physics, the military interest and support still prevalent at the time. I remember during my stay in Bristol, a summer school specifically meant for industrial engineers, a formula which inspired us later in Orsay. Nevill also showed a keen interest in the Reading IUPAP conference on semiconductors which in 1952 brought together physicists from the large industrial concerns as well as university people from all over the world. Nevill was very conscious of the advent of transistors, conscious also of the essential contributions he had made early in the field, indeed before the war, with the analyses of donor states and of rectifying metal–insulator junctions. This interest in the industrial aspects of semiconductors, fostered later by his contacts with Ovshinsky, would lead him to concentrate on the insulator–metal transition of semiconductors with heavy doping; this was also to be an important part of his Nobel work.

Nevill played a decisive role in my life in introducing me to Mary, my wife. This led to many happy visits with Ruth and Nevill in Stuart House, in Sedley Taylor Road, in Caius College, in Aspley Guise and, in recent years, to frequent short letters from him. Besides a keen interest in the life of his family, these letters showed an interest for religion which took an increasing place beside his preoccupation with society and world affairs. It started, I believe, in frequent discussions with Mervyn Stockwood in Cambridge as an intellectual preoccupation: How can one be a scientist *and* a Christian? It developed as a sincere wish to take part in what he thought was a unique human experience and tradition. Faith did not seem to play much of a role in his outlook and I found it difficult to discuss with him.

How best to conclude but to cite Fred Seitz in a recent letter: 'Nevill did much to help Europe redevelop its scientific strength in the post-war years. He had a brilliant career and an enormous number of admirers.'

My Brother-in-Law

Mary Friedel

In 1936 I was recovering from two years of thyroid trouble. After two successful operations, Nevill suggested I should join Ruth and him for a holiday in the north-west of France; it was a never-to-be-forgotten experience. We were to travel in Emily, Nevill's old Austin 10, which barely protected us from the rain. We were to sleep mostly in tents, choosing our ground as we liked, with the permission of the owners. This was just before organised camping-grounds developed with the *congés-payés*.

We started from Calais and went south-west to Chartres on the way to Vendée and Brittany. We stayed three days in Chartres, which Nevill knew from previous visits; he had even made a cardboard model of the cathedral. Nevill and Ruth carried with them a thick book on Mont St Michel and Chartres, which they studied earnestly.

In the country we often had an inquisitive gathering of children to watch those mad English people camping like gypsies. Suspicions were increased when Nevill took water in a saucepan from a village well. Matters were settled with the indignant villagers when a couple of gendarmes asked for our papers and all ended quietly in a discussion on the Spanish war.

I remember vividly a night in my tent at Locmariaquer with the moonlight shining on the sea between tall ghostly black trees. We also found another impressive spot on the long path leading to Mont St Michel, where we spent a night alone, free from visitors.

Emily, Nevill's Austin 10, on the ferry to Saint Nazaire (drawing by Mary Friedel).

We returned through Normandy to Le Havre, where we abandoned Emily on a car ferry. We crossed to Southampton on the *Ile de France*; our somewhat bedraggled clothes and Nevill's beard contrasted strangely with the well-dressed travellers on board.

This was one of the rare holidays Nevill had *en famille* without science.

My Uncle

Jean Friedel

Nevill Mott was one of the Wotans of my personal Valhalla, someone I couldn't even have dreamt of being close to, and yet a very familiar character who very kindly received our family, and later on myself, every year of my childhood. We used to come and live 'around' him and I remember as a little boy he mostly appeared and disappeared in a quiet lofty way, walking up or down the garden path on the tip of his shoes. 'Hello, my dears' would be the words greeting Aunt Ruth, together with my brother Paul and me, if we encountered him on his way back from the Lab. In fact, his silhouette is in my mind definitively associated with the giant multicoloured hollihocks and the astringent August scents of the garden in Sedley Taylor Road at Cambridge. The place in itself was one of the paradises I used to long for as a small Parisian french boy, and one of the rare moments Uncle Nevill could have seemed closer was when he decided he would light a bonfire at the end of it. What a relish it was to watch him fight with the rather damp lawn cuttings, with their slightly fermented smell in my nose and the taste in my mouth of an enormous blue plum down from the tree.

All the same, we used mostly to listen to family conversation at breakfast, or dinner, or supper, as my mother managed to get information out of those two rather silent scientists that were Nevill Mott and my father, his pupil, most often about people they knew or about events in the 'Cavendish'. Uncle Nevill always put his head out, nodded for two seconds with a smile on his face, and looking down at the table, would utter some profound understatement in answer to my mother's rather provocative comments or questions. I think this was a habit they had since they came to know each other, because my mother had always been his quite younger sister-in-law. Aunt Ruth was our direct contact with this world which we instinctively knew was special, and luckily she was all ready to help us get through its sometimes rather awesome aspects.

Later on, I managed to experience a direct relation with Uncle Nevill, especially when he visited us in Paris and during two stays when I went over to England, first alone then with a friend to perfect my English. Although constrained, conversation with him became possible, even if I most often felt like a fool for sharing none of his scientific interests. The difficulty brought in by our mutual shyness was aggravated on my side by a queer dichotomy which made me consider English as a language to speak with women. But it is quite possible that this difficulty to use English with men was in fact a consequence of having such a shy-making uncle. In fact, I got to manage quite well with Uncle Nevill when it was my turn to receive him as an adult with my wife in Strasbourg.

Getting to know, and eventually living next to, such a character has, I am quite sure, profoundly shaped my outlook on world and people, and if I regret not having had more direct and intellectual contacts with the man, gratitude is the persistent feeling I have towards him for having been part of my childhood's rather uncommon background.

PROFESSOR J. FRIEDEL undertook his PhD in solid-state physics with Nevill Mott in Bristol, 1949–1952. He is Emeritus Professor, Université Paris Sud, Orsay. MARY FRIEDEL, née Horder, is Ruth Mott's sister; she married Jacques in 1952. JEAN FRIEDEL, elder son of Mary and Jacques, is a dermatologist in Châlon-sur-Saône, Burgundy.

Memories – with Gratitude

Doris Kuhlmann-Wilsdorf

On 3 January 1949, after a long ride in a closed compartment of a British military train from Germany to Hook van Holland, then a stormy crossing of the Channel and a further train ride, Frank and Margaret Nabarro met me at the Bristol railroad station. On the recommendation of Frank Nabarro, Professor Mott had invited me, to my great surprise, for a three-months research stay at the Physical Laboratory of Bristol University. Civilian traffic between Germany and other countries had not yet been established, hence the ride in a military train, as defeated Germany was still effectively isolated from the outside world. However, the first postwar meeting of the German Society for Metallurgy had taken place in Stuttgart in October of 1948 (in largely destroyed Stuttgart, participants were mostly lodged in the Bahnhof's Hotel, being the air-raid shelter subdivided into tiny cubicles with bunk beds and a communal lavatory). Frank Nabarro, then one of the most senior and long-time coworkers of Professor Mott, had been the greatly honored and only foreign participant at the conference.

There were two reasons that I, then a lowly postdoctoral fellow under my previous supervisor, Professor Georg Masing of Göttingen University, who for lack of official funding provided for my bare needs from his own personal salary, attracted the attention of this honored guest; on account of my American-born mother I spoke much better English than any of the leading personalities, and I had read several of Frank Nabarro's and Mott's papers. Indeed, my PhD thesis dealt with dislocation behavior and as a result, next to Professor A. Kochendörfer of Stuttgart, I was the only person in Germany who could converse somewhat intelligently on the subject of dislocations. The distinguished leaders of the society had therefore informally made me the hostess and companion of Frank for the duration of the conference. As it was, we very quickly established a fine personal accord. This was especially easy because my family and I had been strongly anti-Nazi throughout, and because in the early Hitler years until her emigration in 1937, my best friend, Liesel Stern, was Jewish.

Frank and Margaret Nabarro had invited me to sublet their spare room in their Great George Street apartment, within easy walking distance of the laboratory at Royal Fort. And so, until Frank joined A. H. Cottrell in Birmingham, there followed three months of almost incessant discussions on dislocations. They began over breakfast, continued intermittently in the laboratory, then at lunch (often with Jan and Minnie van der Merwe in the nearby museum cafeteria), and commonly extended after dinner into the night. How fortunate I was to have had the benefit of such incomparably excellent instruction – not that it was free of stress, but certainly most educational.

On the very first morning, Frank Nabarro introduced me to Professor Mott in his large office. Then, as always, he was most kind and considerate and seemed to take a true personal interest in me, as he did in all of his coworkers. In this he was unlike German professors, who tended to be aloof and politely impersonal. The ability of Professor Mott to establish a personal rapport with colleagues and subordinates has doubtless been one of his great strengths. His methods to come to know and judge the scientific abilities and dedication of these were happily agreeable and most effective. Sadly, I have never again had the pleasure of working with someone of Professor Mott's superior gifts in this direction. Most effective were the gatherings of all scientific members of the laboratory for morning and afternoon tea; the gatherings themselves lasted at least thirty minutes and some of the discussions could continue for hours.

Attendance was obligatory. Professor Mott would unobtrusively but effectively ensure that teatime really was a scientific instead of a social gathering by introducing someone after the first few minutes; perhaps he would say, 'Dr Cabrera, you and Dr Burton just told me about your fine discovery on faceting in crystallization from the vapor; please tell us more about it.' Or he might say, 'Dr Eshelby, could you perhaps briefly outline how far you have progressed with developing the elastic theory of dislocations? I should much like to understand this better.' Or to Jacques Friedel, 'I did not quite understand what you told me about the theory of electric conduction in stress fields yesterday.' And to ensure that all were attentively following the arguments instead of letting the discourse wash over them like rain, Professor Mott would interject with questions such as, 'Dr Mitchell, what do you think of that?' Or 'Mr van der Merwe, does this not perhaps conflict with the equilibrium conditions?' And of course Professor Mott, with his superb knowledge of physics, would catch glitches and inconsistencies by making polite but decisive, corrective statements such as, 'But the wave functions would then not be continuous across the interface.'

Another very effective method of honing the scientific level of the laboratory was selective informal conferences in Professor Mott's office. I vividly remember one, including only three or four of us besides Professor Mott, in which Mr Baird, a Canadian graduate student, was to report on the progress of his PhD thesis on the theory of track formation by charged particles in photographic emulsions. This was very far from my own area of expertise. I imagine that Professor Mott included me simply in order to gauge my general physics background, since I had graduated in physical metallurgy, and perhaps gauge my ability 'to think on my feet'. For the same reason Professor Mott presumably often included 'outsiders' in such conferences. In this particular case I was very lucky, as he suddenly turned and asked me, 'Miss Kuhlmann, what do you think of this?' I could correctly say something like, 'Well, there is a flaw here which needs attention. The second last equation at the bottom of the blackboard traces the functional dependence on the blackening to high speeds while at the outset a small velocity was assumed and the higher terms were dropped.' More typically, such small conferences were among those working on closely related topics. Thus Professor Mott might ask Charles Frank, Frank Nabarro and me for a conference in his office on C. F. Frank's theory of dislocation multiplication by reflection, which was then very popular. I always strongly, although perhaps not very diplomatically, disagreed, since for my PhD thesis I had studied slow aftereffect creep on loading of metals to below their macroscopic yield point, and from this was unshakably convinced that dislocations had the ability to move and multiply slowly and continuously.

Hitler's Germany had felt like a mousetrap to me, and for years I had a fervent desire to emigrate. In the prevailing and long-lasting isolation of Germany this was evidently very difficult. While my fiancé, Heinz Wilsdorf, and I planned ultimately to go to the United States, as a first step I needed to stay in England. Therefore I begged Frank Nabarro to put in a good word for me so that Professor Mott would invite me to stay permanently. I was overjoyed that Professor Mott did indeed make that offer, partly also on the strength of my very first scientific publication, even before my PhD thesis and somewhat disconnected from it, namely the theory of aftereffects.

In the first three months of my officially temporary visit and in the following year, until my marriage in January of 1950 and our emigration to South Africa four months later, Professor Mott repeatedly invited me to his home. It was a little house across the wide courtyard of the Royal Fort. Thus I came to know and cherish his family, in particular Mrs Mott who was universally loved and admired, and their delightful daughter Alice. Most of those invitations were for large parties, including many members of the laboratory and other members of the university. How very different they were from any social gatherings I had known! In Germany guests are seated, typically around a table, but here guests stood with drinks in their hands and were offered refreshments from trays! Professor Mott was a very gracious host and introduced the guests to each other. I was overwhelmed by the many different people whom I had never met, and I felt very awkward because of a lifelong inability to recognize people and remember names. (I'm glad to say, I can remember my close family!) Thus for me it was somewhat unnerving when Professor Mott would interrupt conversations which had just started to go smoothly with 'Miss Kuhlmann, I am sure you will enjoy meeting ....'. I thought he might be trying to relieve my conversation partners after I had somehow behaved improperly. Only at the second or third party did I realize that, as a perfect host, Professor Mott wanted to make his guests comfortable and facilitate mutual exchanges.

At the very first party, I met Dr Elsie Briggs, University Appointments Officer. I fell in love with her on first sight, and being repeatedly separated from her by way of new introductions, I again and again made my way back to her. Dr Briggs became one of my dearest friends to whom I turned in all kinds of troubles and always left consoled and with good advice. Soon she managed to get my future husband to come to England on her personal invitation. At our wedding in January 1950 in London she was our witness, and both of us have cherished her and kept up contacts until her death several years ago. Enduring relations also developed with Mrs Terry, one of Professor Mott's two secretaries. Most kindly, soon after my arrival, she gave me a few dresses as I had but little clothing left after many years of scarcity, war and occupation. She and Mrs Langdon, Professor Mott's other secretary, were wonderfully kind people. Also very important were enduring friendships established among coworkers, not only with Frank Nabarro but also with Jan and Minnie van der Merwe. And Dr Eshelby visited us in America, just when I had to start my first lecture course on dislocations at the University of Pennsylvania. What would I have done without his crash course on equilibrium equations, compatibility equations and the Airy stress function in between meals and diaper changes of our son! Sadly he died fairly soon afterwards. Professor Mott made all of these friendships possible.

On a few other occasions Professor Mott invited me personally for lunch or afternoon tea with his family. The last was shortly before my final departure to South Africa. Professor Mott regretted my leaving, as did I, but I had no choice since I was now married and my husband had been unable to find a position in England, in spite of Professor Mott's efforts. As we also had failed to obtain a visa for the United States, my husband had already left for a good position in South Africa, through the assistance of Jan van der Merwe. As a farewell, Professor Mott invited me to a family lunch, I believe it was a Sunday and the birthday of Alice. Knowing that I would visit my mother-in-law in the eastern sector of Berlin before sailing for South Africa, he took my firm promise that I would not enter the Eastern Zone. He feared that I might be detained on some pretext, especially since I had worked with him in Bristol and Soviets might think me a spy or else to have some useful information. This was characteristic of the kind and considerate person that he was. My final memory of Professor Mott in Bristol is of blowing up balloons for Alice's party that afternoon. The balloons were unexpectedly tough and all of us were straining as we blew, oohing and aahing with every success.

In later years Professor Mott visited us twice in the United States, once including long discussions about the theory of work hardening. I cherish the memory of these brief visits, but the most vivid pictures of him remain from my days in Bristol. Among his charming personal traits that I have always valued, one especially sticks in my mind. During informal conferences in his office Professor Mott might slowly pull up and discard one or two sleeveless pullovers, and perhaps put one of them back on, meanwhile never missing a beat in the discussions. Truly, knowing Professor Mott and learning from him as a person and as a scientist has been one of the great gifts of fate. He has had an incalculable and very beneficial impact on my life and on my husband's.

Nevill Mott: Reminiscences and Appreciations

PROFESSOR D. KUHLMANN-WILSDORF was born in Bremen, Germany, in 1922 and after completing an apprenticeship as metallographer and materials tester at the Focke-Wulf aircraft company in Bremen, studied physical metallurgy under Professor Georg Masing at Göttingen University from 1942 to 1947. She continued as a postdoctoral fellow with Professor Masing until her stay in Professor Mott's laboratory from January 1949 to May 1950. She then followed her husband to South Africa, where she served as lecturer in physics at the University of the Witwatersrand in Johannesburg, for her last three years under the Departmental Chairmanship of F.R.N. Nabarro. In 1956 she joined the Department of Metallurgical Engineering of the University of Pennsylvania in Philadelphia. For the past three decades she has occupied the position of University Professor of Applied Science with dual affiliation in the Department of Physics and the Department of Materials Science and Engineering at the University of Virginia.

Recollections of Bristol 1951–52

Alfred Seeger

'Glück ist, wenn einer immer da steht, wo er hingehört.' (Happiness is when one always stands where he belongs.) If we adopt this characterization of happiness by Theodor Fontane (1819–1898), the German writer, poet and reporter, Nevill Mott was a happy man. This is certainly the way I shall remember him.

Let me go back for a moment to the years following the Second World War. The war had caused me to lose almost two and a half years of my formal education, but I managed under rather trying circumstances to read physics at what was then the Technische Hochschule Stuttgart from 1946 onwards and to obtain a doctoral degree early in 1951. During this period I became strongly aware of the isolation from the outside world that had developed in Germany since 1933 in every respect – cultural, scientific and social. It was clear to me that I should attempt to work in a foreign country as soon as possible and to familiarize myself with what we called the 'outside world'. In those days the opportunities for young Germans to do so were extremely rare; the main exceptions were scholarships offered by the British Council. When I applied for one of them, there was no question of whom to name as possible host. The physics background I had acquired in my university years was mainly in solid-state physics, so Nevill Mott, at that time still in Bristol, was the obvious choice.

The biggest problem I had to face was that I was several years short of 25, the minimum application age. Somehow I persuaded the rather reluctant pre-selection committee to pass my application on to Bristol. In retrospect, the day the acceptance letter arrived, in early 1951, appears to be the most important day in my professional life. I never talked to Nevill Mott on how, knowing only what I had written, he had persuaded the British Council to break its own rules. Perhaps he thought of his own unconventional career, having at 27 been appointed the first Professor of Theoretical Physics at Bristol University, before he had time to obtain his PhD, and having been preferred on this occasion to perhaps more obvious older candidates.

Be that as it may, I arrived in Bristol on 1 October 1951 and met Nevill Mott for the first time. I was immediately impressed by him. Here was a man who obviously stood where he belonged. As departmental head he directed, with an easy hand, what was, at least in my judgement, the most esteemed physics department on this side of the Atlantic, leading in both solid-state and elementary particle physics. He was himself active and successful in several solid-state research fields, among them the question of why NiO is an insulator (this led eventually to his Nobel-prize winning work), the electron theory of metals (in particular the so-called screening problem, from which the work of G. E. Lee-Whiting, J. S. Plaskett, J. Friedel, F. Fumi and others emerged), and dislocation theory with applications to such down-to-earth problems as work hardening. He ran his 'own' journal, the *Philosophical Magazine*, and was just about to launch another one, *Advances in Physics*. These journals enabled him to get papers quickly published that he thought should be read by the scientific community, even when they came from unknown young scientists and/or contained unconventional ideas. Nowadays, when we have too many science journals and when almost anything can get published if one shops around long enough for a lame journal that eventually publishes what has been written, it is hard to imagine how important this was, particularly for young scientists. Nevill Mott had gathered around him and the other leading members of the Bristol solid-state group young theoretical physicists from all over the world, or more precisely, from the Commonwealth, plus a few Europeans and an occasional Japanese visitor. Most of them had come to work for a PhD. Nevill and Ruth Mott provided the social atmosphere for this rather heterogeneous group and brought us in contact with interesting people outside the department.

The Mott family lived in the Royal Fort House next to the Physics Department. After the standard working hours they often invited some of us youngsters into their house for a 'glass of sherry' and for discussions on a certain scientific problem. The sherry was real but the 'scientific' discussion usually soon turned into one on art, music, politics, economics, linguistics, sociology or something else. In retrospect, I now think that, to some extent, the science was just a pretext and the main idea of Nevill and Ruth was to bring together young people with very different backgrounds and let them exchange their views freely. On these and other occasions I became familiar with the enormous breadth of Nevill's interests and background knowledge. He took a particular interest in modern languages and was at the time the president of an institution which was called (if my memory serves me right) the Modern Languages Association. Its aim was the furthering of the understanding between the European nations by advancing the learning and understanding of foreign languages.

One episode is characteristic of Nevill Mott in his last Bristol years. He had been invited to Italy to give a lecture on the current research in his laboratory. His hosts had told him in advance that giving the lecture in English might present a problem for his audience but that French would be understood without difficulties. So Nevill prepared his lecture in French, only to find out that in this language he was not able to reach his audience either. He told his hosts that he would go back to Bristol, learn Italian and come back to give the lecture in Italian. This is what he did, though he never told me whether it worked.

Another time that Nevill Mott liked to spend with his young guests was from eight to nine in the morning, before the rest of the laboratory sprang to life. I remember that he frequently asked me to come to his office early next morning. Apparently he liked to use me as a guinea pig for new ideas he pondered. In those days my way of doing physics was quite different from his – maybe this was the reason why he wanted to know my opinion. Before I came to Bristol I had been strongly influenced by the Sommerfeld school of theoretical physics, with which Nevill Mott had been in touch in his early career, too. Its main characteristic, the mathematical deduction of physical conclusions from preset models, was virtually orthogonal to Nevill's way of doing physics after he had moved to Bristol in 1933. He had not only a tremendous intuitive feeling for the essential physics involved but almost always found a simple way – often no more than the notorious back-of-an-envelope calculation – to find out quickly whether his ideas had a chance to agree quantitatively with the experimental facts. I like to think that it was this difference in our way of arguing on the one hand and my willingness to follow his thinking that led him to spend so many of these early mornings with me.

For me there is no doubt that, through these and later discussions, Nevill Mott has influenced my way of doing physics more than anyone else. It is to him that I owe my subsequent ability to combine the Sommerfeld-style theoretical physics and the Mott-style intuitive physics with its close contact between experiment and theory. After having returned to Germany this helped me to bridge the gap between experimental and theoretical physics, which was then very wide and has not yet been completely closed. Nevill appears to have approved of the way I developed during the year in Bristol, since two years later, when he had decided to accept the Cavendish Professorship in Cambridge, he invited me to the Cavendish Laboratory to give a course of lectures on dislocation theory and crystal plasticity. A lifelong friendship developed from this, but that is another story.

To end these few recollections, I return to the Fontane quotation at the beginning. I selected it because I believe that, beyond the tremendous impact he made on the physics of this century, Nevill Mott epitomized Fontane's happy man. When in later years we discussed philosophical questions such as, 'What is important in life?' I had the impression that Nevill felt this also. Maybe this feeling was at the root of his interest in religion. He was highly gifted in many respects, and he had the good fortune that virtually throughout his entire professional life, i.e. for two-thirds of a century, he could make full use of these gifts. I think it was the happiness that flew from this that attracted so many young (and now not so young) disciples to him. We have many reasons to be thankful for Nevill Mott, not only Nevill the physicist but also Nevill the happy man.

PROFESSOR DR A. SEEGER was born in 1927 in Stuttgart. He graduated and received his doctorate from the University of Stuttgart, became a lecturer in 1954 and has been Professor of Solid-State Physics there since 1959. He was a British Council Scholar in Bristol in 1951–52 and has held visiting appointments at Cambridge and many other universities. He was Director of the Physics Department of the Max-Planck-Institut für Metallforschung from 1965 to 1995. His research interests are in defects in solids, crystal plasticity, soliton theory, electron theory of solids, and nuclear techniques in condensed matter physics.

A Brief Reminiscence

Robert Cahn

When one is just starting out on one's life's work, it is good to stand in awe of those giants whose achievements one has heard of in one's teens. Nevill was a man one could simultaneously stand in awe of and converse with amicably as a fellow scientist. Not only for solid-state physicists the world over but also for physical metallurgists (of whom there are a lot) he was a unique role model, never to be emulated but nevertheless uniquely approachable. Not only that, but he was also quite prepared to regard metallurgists as approximating to human status.

I first saw him at the end of my student days in 1947, stalking (I thought) rather like a dignified stork back and forth behind the speaker's bench in a large lecture theatre at the Royal Fort in Bristol, as he opened one of the early conferences on dislocations. At the Cavendish not long before, I had observed the comportment of other greats of the world of physics: Pauli nodding his head like an urgent waterfowl; Bohr muttering intensely; Orowan putting things into imaginary buckets and taking them out again, never quite touching his fingers together so that all the onlookers waited breathlessly for the conflagration to be expected at first contact. None of them gave quite the instant sense of a scientific hero that I recall from my first experience of Nevill Mott.

Those Bristol conferences were unlike any others. Mott, Frank and Nabarro made an unforgettable threesome. My first account of the phenomenon of polygonization apparently made some impact there. About a year after the 1947 conference, Nevill wrote to my boss at Harwell and suggested that this young man Cahn might fit in rather well in the Bristol Department; what did the boss think? The boss made it exceedingly plain that this was not to be thought of until I had finished my allotted task at Harwell; so that was that. All of us have had potential turning-points in our lives that never became actual, and this was one of two that I could never quite get out of my mind. Imagine being able every day to talk with Nevill in the tearoom, that quintessential focus of a researcher's day in a British university, about one's latest obsession! It was not to be. But the magic of Nevill, both in actuality and, later, the memory of his personality after he had left, drew me back on visit after visit to Bristol over many years.

In 1953, he became the founding editor of *Advances in Physics*, the counterpart on this side of the Atlantic to those pervasive brown volumes of *Solid State Physics*. Thereupon, he eloquently urged me to write an overview of twinning in crystals for his new journal, and thereby did me a great favour by showing me the possibilities of this kind of writing. He was a most persuasive editor.

Some years after he moved to Cambridge, during the early sixties, I paid him a visit in the Master's Lodge at Gonville and Caius, and he was bubbling over with the architectural marvel of the new, modernistic college building at Harvey's Court. He insisted on taking me there and then to see it while he explained its subtleties. It was a thousand pities that this connection, which clearly meant so much to him, ended in tears.

I again encountered Nevill some time after he had become a member of the academic advisory board of the new University of Sussex. It became clear that he had spoken for me when a professor of materials science was needed there, and with his encouragement I was launched on the creation of Britain's first department devoted to this new disciplinary concept, of which physics was a very major component. Again, it was quite clear that he regarded materials scientists as respectable human beings, at least potentially so.

After that, our meetings were more spaced out until I came to live in Cambridge on retirement. Nevill once again offered an invaluable object lesson in how one should configure one's notional retirement. Another lesson came through the symposium he organized at the Royal Society in 1979 on the beginnings of solid-state physics, which proved most valuable to anyone becoming interested in the emergence of materials science. Not long before his death he gave a beguiling lecture to the Cambridge Philosophical Society about his student days in Cambridge; he had found a cache of letters to his mother which opened the floodgates of memory. He perceived the past through the eyes of the present, a most Whiggish approach. Nevill was a man who could fruitfully look back without for a moment ceasing to look forward.

Though I never had the immeasurable privilege of working under Nevill's benevolent eye, he had as much influence on me as any scientist whom I ever met.

PROFESSOR R. W. CAHN, FRS, is a physical metallurgist turned materials scientist; in certain moods he calls himself a metal physicist. He has professed these avocations in a number of universities – longest at Sussex University – and has returned in retirement to his alma mater, Cambridge. There he devotes himself to the writing and editing of books, encyclopaedias and journals, and also to science journalism in the pages of Nature.

Mott and the Cosmic Radiation

Richard Dalitz

In the spring of 1948, I was a research student at Cambridge University. The small scholarship (prewar level) which brought me there in 1946 from my home in Australia was almost exhausted and my applications for postwar Australian scholarships had been turned down. Being married and with a son, I urgently needed support in order to complete my PhD thesis. Then one day, my supervisor mentioned that Professor Mott had been asking around at Cambridge, seeking to find a PhD student to work with him at Bristol University on high-energy nuclear physics. I wrote to Mott expressing my interest in this position and he summoned me to meet him in London.

Professor Mott told me his opinion that the most important research work in his department at Bristol was the study of the cosmic radiation going on in its fourth floor, i.e. the attic floor of the Royal Fort, the building in which most of the physicists at Bristol University worked. This floor housed the Emulsion Group of Professor C. F. Powell, whom I had heard speak at the Cavendish Laboratory in the early winter of 1947. He talked about his group's discovery of the 'nuclear force mesotron', first envisaged by Yukawa, which we know today as the pion. Mott wanted to move back into this exciting area of nuclear physics, feeling that at Cambridge he had allowed himself to be diverted away from his early work on nuclear physics by the needs of others, especially his fellow research students, who sought his advice and supervision in their studies of problems in solid-state physics.

To achieve this aim, he said, he needed a research assistant who would help him to formulate his ideas and make calculations relevant to the new data from the fourth floor. In this post I would receive a university grant equal to the DSIR grant at that time and give some lectures on dynamics to the first-year undergraduates, which would help me and my family to survive. However, thinking aloud, he expressed some unhappiness that he had been unable to find any Briton for this post, because the students from abroad whom he had appointed to assistantships usually returned to their own country in the end. Why could he never find a UK student? So the matter was left open.

Several weeks later, I received a letter from him offering me this post and I accepted it, arranging to arrive at Bristol at the end of August. While we were looking for housing in Bristol, and I was settling down to my desk in the house which the Physics Department had then as an annexe for theoreticians, just south of the lawn at the southern end of the Royal Fort, my wife was sent to hospital urgently because of internal bleeding. En route there by taxi, she stopped briefly at this house to tell me the situation, leaving me literally 'holding the baby'. As I was standing there with my son, Professor Mott came out of the Royal Fort, so I told him that I would find it difficult to carry out my duties during the next few weeks, asking him to excuse me. He was very kind and told me not to worry, for my wife was in good hands. Later in the week, Emeritus Professor Tyndall, the previous department head asked me to come over to his house for an evening meal with his family. It was clear that I had landed in a happy department.

Term had started before Mott and I had our first discussion. His proposal was that we should calculate the characteristics of the cosmic ray showers due to the collision of a high-energy proton with some target nucleus, which might be ^{12}C, ^{16}O, ^{80}Br or ^{108}Ag, the dominant nuclear species in emulsion. The secondary particles from this collision would generally have high energies, and each of them would have to be followed until it collided with another nucleon in the same nucleus, leading to tertiary particles, again mostly of high energy, and so on. This cascade would have to be followed until each particle escaped from the nucleus or had so low an energy that it remained in the nucleus. These calculations could not be carried out analytically; numerical calculations would be necessary, following the history of each particle, and they would be immense, simply because of the large number of particles resulting from this cascade of interactions.

Powell's group had observed and measured many such showers in emulsions exposed to the incident cosmic radiation at mountain altitudes, and we wondered what we could learn about the fundamental particle–particle forces by comparing our shower calculations with Powell's data. Before we really got to grips with these ideas, Mott's secretary called him out to settle some very urgent problem which had just arisen, and Mott had to depart, promising to be in touch as soon as possible. This was the pattern of our few meetings. We never managed to get together long enough to discuss these cosmic ray problems in any detailed way. This was probably to the good. The real problem with our proposed calculations was that we did not know the high-energy nucleon–nucleon cross-sections, including those for meson production, and the meson–nucleon cross-sections. Indeed the pion, the most frequent meson by far, had been identified first at Bristol, but only late in the preceding year. All that could be done at this stage was to use cross-sections resulting from untested theories, and that was not an attractive proposition.

So the weeks passed. I soon came to know Jean Heidmann from the Leprince-Ringuet group at Paris, whom Mott had asked me to keep an eye on, and K. McClure Baird, a physicist in optics from the National Research Council at Ottawa, the three of us being in the same room, with Jock (John) Eshelby next door. Most important for me, I met all the fourth-floor research students, especially Ugo Camerini, Hugh Muirhead, Owen Lock and David King, and had many discussions with them about their work and their data. I recollect that the first 'tau-meson' event, a K^+ meson decaying to $\pi^+\pi^+\pi^-$, was discovered at Bristol about the time of my arrival. This made me aware of it from the very beginning, which made it natural for me to undertake my later studies of this 'tau' decay mode and its competing 'theta' decay mode, $K^+ \to \pi^+\pi^0$, which led to the questioning of parity conservation in these decays and ultimately to the realisation that all weak interactions have parity-violating parts.

My PhD thesis research for Cambridge University needed a lot of my time as well, and Professor Mott had me give a theory seminar about it under the title 'Zero–zero Transitions in Nuclei'; these transitions involve electron-positron pair creation. Many questions were raised by Mott and by the audience, which I found very helpful. Now and then, Mott would send me a paper to referee for *Phil. Mag.*, the local journal, which he edited. I regularly gave my lectures on dynamics, as arranged. In his relationship with the Bristol undergraduates, Mott was ahead of his time. He encouraged them to make their grievances known to him, at open meetings. I remember one meeting about the teaching of mathematics to physics students, which I made a point of attending, since I was lecturing on applied maths. The students felt that this had been taught to them in an 'old-fashioned' way – using Cartesian coordinates rather than vectors. I was able to tell them that this was not the case in 1948–49 and the first-year undergraduates did support me on this. Those complaining were from the second year and knew nothing of me. Several of those first-year students became my good friends in later life.

The Bristol department had a wonderfully friendly atmosphere, which was such a great contrast with Cambridge. Mott was the major cause of this. He and Mrs Mott often held evenings for research students, young postdocs and faculty, and their wives; their front door opened to the lawn just south of the Royal Fort. The invitation was always for coffee but I don't recall ever having coffee, it was always wine. We sat around the room, facing the coal fire, which we all enjoyed in those days. About every half-hour, Mott would move his guests from one seat to another, so the group would get to know one another more quickly, which I thought a very effective procedure. However, I do recall one occasion when Mott wanted to exchange a wife sitting by the fireside with another wife near the door; the woman by the fireside flatly refused to move, 'I don't want to move. I am so happy with my present seat.' Mott was nonplussed and went on to make other exchanges, not returning to the woman for some time. Our recollection is that she defied him successfully.

Mott still had our project on his mind, of course, and sent me to visit the Dublin Institute of Advanced Studies to consult with Professors W. Heitler and J. Janossy, who were calculating meson production cross-sections. This was agreed with the Dublin IAS and I spent a week there. I learned that Heitler and Janossy were just completing a long paper on the properties of cosmic ray showers, based on their calculated cross-sections, and that their main interest in my visit was to learn as much as possible from me about the current Bristol measurements. They had not mentioned this in our correspondence, and I had not brought any new data with me – Powell would not have allowed it, I am sure – since my interest was to learn the most fruitful methods of calculation. All that I could recall from discussions on the fourth floor was consistent with what they knew, and they were satisfied with that, but they were not inclined to spend much of their time discussing their calculations with me. Their deep interest in the Bristol data confirmed for me its great significance at that time. In essence, they had already done what we were proposing to do. In the long run, however, their calculations had one serious defect; they allowed at most one meson to be produced in any nucleon–nucleon interaction. This restriction became a matter of principle with Heitler until hydrogen bubble chamber observations showed that, at sufficiently high energies, proton–proton collisions readily gave rise to the direct emission of two or more mesons. I had an interesting session with Professor Schrödinger, and met the young DIAS theoreticians Corinaldesi, Ning Hu, Basu, Symonds, Field and S. N. Gupta, each coming from a different country.

In 1949 I moved to a long-term appointment at Birmingham University. An early study I made there concerned Mott's cross-section for the scattering of an electron by a Coulomb field due to a charge Ze which he had derived in 1923. For small Z, i.e. light nuclei, the approximation

$$\sigma_{\text{Mott}} = \sigma_{\text{Ruth}}\{1 + Ze^2\Phi(\theta) + o(Ze^2)^2\}$$

has often been used and Mott gave an expression for $\Phi(\theta)$. I noticed another expression for $\Phi(\theta)$ used in the literature. Which was correct? I used the new Feynman graph method as a check, but I was surprised to obtain a third expression for $\Phi(\theta)$. I went back to Mott's original paper; the crucial function was a sum of a function $c(n)$ over integers, from $n = 0$ to infinity. The catch was that the term $c(0)$ needed special treatment, and was not given by $c(n = 0)$. Correction of this term in Mott's paper led to my result for Φ. I wrote to Mott, asking whether he still had his original calculations. He replied to say that he was sorry about his error, but all of that was so long ago it could not be checked. With A. Alexandrov he recently published his selected scientific papers accompanied by a commentary. I see that he didn't remember this point in his comments, but it is of no importance. Today, with computers, everyone uses the complete expression, not the approximation.

I did not see Mott again for many years, mainly because I was working in America. After my return to a post at Oxford in 1963, our paths crossed occasionally but briefly. Our longest conversation was at Sir Charles Frank's eightieth birthday meeting on 15 April 1991, in the new tearoom of the Royal Fort. Mott greeted me with, 'You seem to have found a lot to say for yourself, Dalitz,' and we sat down to talk, rather intensely. I told him that I had no feeling of guilt in my failure to guide him back into high-energy nuclear physics, since other fields of great importance had also begun to develop at Bristol during 1948–49. For example, I recalled the morning when Charles Frank came in to coffee and told us how important dislocations were in the growth of crystals, showing us how the screw dislocations generate 'whiskers' growing out from the crystal surface. Mott's conversation moved on to other topics and I asked him for his recollections of the Physics Club, a group of ex-Cambridge physicists who had taken up senior posts in other universities, and who met on one Saturday in each term, generally in London. He replied that he did not know of any such club.* Finally, he spoke of Klaus Fuchs (this suggested to me that he had forgotten I was from Australia, all of my fore-fathers having emigrated from Europe in the 1850s) and mentioned a lecture which Fuchs had given to mark the centenary of the birth of Niels Bohr, saying 'Fuchs said that Bohr had advocated sharing the secrets of the atomic bombs with the Russians, while he, Fuchs, had given the secrets to them.' When I mentioned this to Rudolf Peierls, he was very interested to know what Fuchs had said, so I wrote to Mott to ask for the reference. After some time, he replied (the letter is undated):

> Dear Professor Dalitz,
> Some time ago you asked me about Klaus Fuchs' essay for Bohr's 100th birthday – I couldn't find it. It has now come to light and if you are still interested I will send it. Let me know 63 Mt. Pleasant, Aspley Guise Milton Keynes MK17 8JX – or telephone 0908 5832
> Yours sincerely,
> Nevill Mott

The preprint from Fuchs bore the greetings, 'Sir Nevill Mott/in gratitude/Klaus Fuchs/29.9.86.' When we read the address, we found that Fuchs had not mentioned his part in the Manhattan Project or his providing the USSR with plans, reports and materials from that project. It was sometimes easy to misunderstand what Professor Mott said. I could give several other examples of this kind.

I sought to bring our discussion back to my interview in 1948, but I then realized that we were the only persons still in the tearoom. Since Professor Mott was a speaker in the next session, we got up and talked no more.

That subject stuck in my mind for several years and I felt the desire to complete the discussion. Finally, in 1995, I wrote him a letter in which I said that my year at Bristol had been of vital importance in my life, partly because it enabled me to recover from the effects of Cambridge and to meet a much wider range of physicists than would have been possible anywhere else. I said again that I felt no guilt in my failure to bring him back into particle physics. There were too many people urgently needing his advice on questions to do with solid-state physics and he was under great pressure from them. Besides,

---

* It was interesting to read recently in *Notes & Records Roy. Soc. Lond.*, **46**, 171 (1992) that Professor Mott was Chairman of the Physics Club for some period, up to the end of its fifty-sixth meeting, held in the autumn of 1947, when he had exceptionally invited it to meet at Bristol. Of course, Mott was involved in so very many activities during his lifetime.

he had done so very well in that field, deservedly gaining a Nobel prize for his work. My letter ended:

> What did I want to say to you? I suppose I just wanted to remind you of that interview [in London] and to say 'Well, here I am still after all, you see.' I do have regrets that I did not get back to my native Australia, of course.

He replied on 6 December:

> Dear Professor Dalitz
> I don't think I ever answered your letter of 8 Nov. I am glad you have such good memories of Bristol. So have I – and particularly of Arthur Tyndall.
> What fascinates me in particle physics is how the results from the big machines are used in exploring the first few minutes after the big bang.
> But solid state physics remains very satisfying. My present interest is high-T_c superconductors, which I think we understand, though we don't agree with Phil Anderson and most of the Americans.
> With best wishes
> Nevill Mott.

Professor Sir Nevill Mott was a most remarkable man, no matter from which view you looked at him.

PROFESSOR RICHARD DALITZ is an Emeritus Professor of Physics at Oxford University, working in its Sub-department of Theoretical Physics since his retirement in 1990 as a Royal Society Research Professor, a position he held from 1963. He was educated in Melbourne, Australia, and obtained his PhD (Cantab.) degree in 1950 after work at Trinity College, Cambridge, and the universities of Bristol and Birmingham. His research has been concerned with various aspects of nuclear and particle physics. His active interest is with the properties of the top quark.

Reminiscences of Bristol

Charles Crussard

I made the acquaintance of Nevill Mott in 1947, when my friend Pierre Coheur persuaded me that the must for metal physicists was Bristol and the summer school organised there by Professor Mott. We decided to attend.

Of Nevill Mott I remember first of all a smile and a voice; a loud voice elegantly giving the heartiest welcome to his guests, and explaining the most difficult matters of metal physics in perfect English, easily understandable by foreigners.

He introduced the summer school brilliantly, recording the pioneer works of Polanyi, Orowan, G. I. Taylor and J. M. Burges on dislocations, and exploring how they interacted with internal stresses. Frank Nabarro frightened us with his avalanches of dislocations! Alan Cottrell reassured us by explaining how the latter could be slowed down or stopped by impurity clouds. Robert Cahn explained how arrays of dislocations could constitute walls perpendicular to the glide direction; an observation which Egon Orowan generalised a few months later in Paris, at the Journées d'Automme of the French Society of Metallurgy, explaining many subgrains in metals by what he then called polygonisation. As a matter of fact, I had observed it a few years before and called it recrystallisation *in situ*. In his conclusions, Professor Mott predicted great success for this new theory of dislocations.

In 1948 I again met Nevill Mott at a memorable meeting in Amsterdam, where he had to face a very aggresive Linus Pauling! Writing on two mobile blackboards, which he pushed like shields against Mott sitting in the front row, Pauling pretended he could very well explain conduction by a resonance mechanism of valence electrons, and challenged Mott to localise sufficiently his band electrons in the case of non-metallic materials. Mott's loud voice was heard, explaining calmly how he could do it.

I returned to Bristol in 1949. Professor Mott and his colleagues had improved their mastery of dislocations: avalanches would soon be replaced by quieter sources. The most striking event of the meeting was when Charles Frank was explaining the mechanism of growth spirals, and J. I. Griffin rose from the floor and related that he had actually observed such spirals on beryl. Later we were a bit disappointed to learn that the event had actually taken place a few hours earlier in a smaller circle!

Jacques Friedel attended this meeting. Thanks to the closer links he established later, I had more opportunity to meet Nevill Mott in a nice and friendly way, and to appreciate his outstanding personality.

Nevill Mott: Reminiscences and Appreciations

PROFESSOR C. CRUSSARD's *career has been devoted to research on metals. Graduating from the Ecole Polytechnique and from the School of Mines, Paris, he was introduced to metallurgical research by Pierre Chevenard in Imphy, where he spent a year. After joining the School of Mines, he became the first director of the newly created Metallurgical Research Centre, and later Professor of General Metallurgy. He was called by the government of India to establish the programme of the newly built National Metallurgical Research Centre of Jamshedpur, which he directed for six months. In 1952 he joined IRSID, the French equivalent of BISRA, where he started and directed for some years the Physics Department, and became Research Director. In 1963 he joined Pechiney, the largest French company for non-ferrous metals, where he was Scientific Director.*

Mind Without Frontiers

Edward Steen

When I first met Nevill, my wife's cousin by marriage, it was before journalists moved, as now, straight from being opinionated teenagers into the opinion pages. For me there were the traditional, union-ordained humiliations of cub reporting on a local newspaper. Village fetes were a routine ordeal, but there were many other, exquisite forms of revenge on the graddies. Another favourite was to assign us to cover parish councils in the evenings, in commuter villages where nothing, but absolutely nothing happened. 'You wouldn't know a story if it came up and bit you,' the news editor would say triumphantly next morning as he spiked the copy. And there were vox pops in the rain, though perversely I came to enjoy them.

One Saturday, fresh from honing my adjectives on a dozen simultaneous five-a-side football matches, I came to fetch Sally, my wife, and our small daughter, Lisa, from Sedley Taylor Road. Nevill was standing in the garden, it was a beautiful spring day. 'Did they win?' he asked loftily. It could easily have been taken for arrogance, or even a special 1930s brand of cruelty. But it felt like, and was, something quite different – an expression of fellow feeling which was hugely flattering. Perhaps that's why I remember this moment so well.

It was curious, this sympathy between us. It had nothing to do with science. Perhaps it had a little to do with both being very tall. Later, when he was in his eighties and I had at last escaped to Fleet Street, he told me about the inducements he was being offered to attend international conferences. He had received a letter which urged him to come to Colorado, since he could fit in several days of ski-ing. 'Grand Slalom' was a snippet of absurdity that I smuggled into the august pink pages of the *Financial Times*.

Ordinary people, of whom I count myself one, often find great intelligence daunting. Nevill, if he was a genius, was never at all frightening in that way. When our children were small he would bring out a huge box of worn brown wooden building bricks, of all shapes and sizes, for them to play with. And he would smile down on them. Something transmitted itself across the generations. Nor was his *bonne volonté* something abstract. During that time in Cambridge, through some misfortune or incompetence, probably both, we needed somewhere to stay. And we were welcomed without question to the airy house in Sedley Taylor Road. It sometimes reminded me of the largeness of spirit, conviviality and the aura of kindness which Ben Jonson described in his poem 'Penshurst.'

I remember marvellous conversations about anything and everything; in retrospect they had considerable influence on me, though they never had the character of teacher and pupil. Particularly his knowledge of, and interest in, Russia and Eastern Europe inspired me greatly later on. We also spoke sometimes of Clifton College, where we had both been at school, and which we both found enervating in similar ways. I never quite understood why Nevill maintained a connection there. Perhaps we felt he could change things for the better, and he probably did.

I discovered only later how practical were his ideas on educational policy, and how foolish the politicians not to take more notice of him in this respect, or to recognise his far-sightedness on nuclear energy. Against stupidity even the gods strive in vain, and Nevill never seemed to be bitter on this score or on any other. What I admired most in him was that dryly humorous perspective ('Did they win?') which was only partly a function of having seen almost the whole peculiar century. It went naturally with his internationalism. He recognised connections, and strode over cultural or ideological boundaries without hesitation. Behind the most transparent of masks he was, eventually, that most unusual phenomenon, a good man. All of us who knew him were the better for it – we all won something.

EDWARD G. STEEN, born 1950, was educated at Innsbruck and Besançon universities, and read English at Clare College, Cambridge. He wrote for the Financial Times *before specialising in foreign affairs, especially Eastern Europe, at the* Telegraph *and the* Independent, *which he helped to establish in 1986. He is currently an official with the Brussels-based federation of Christian Democratic political parties in Europe.*

A Brilliant Shaft of Light

Frederick Seitz

My introduction to Nevill Mott occurred through the scientific literature when a beginning student. I had been studying quantum mechanics and, somewhat naively, thought of the new laws of quantum statistics as principles to be applied primarily to statistical ensembles. Then, on reading Mott's paper on the scattering of identical particles, I suddenly realized that the laws applied in dynamic as well as static settings, an upgrading of vision. Soon thereafter, Mott published, with Massey, an entire book on scattering processes. I joined with a group in going through the volume in a series of seminar meetings which further broadened our perspectives.

Solid-State Physics

Eugene Wigner and I had, by good fortune, made progress in the problem of developing realistic wavefunctions in simple solids, starting with metallic sodium. Since this work opened up a new area for research, I decided to make it my principal interest in the postdoctoral period, starting in 1934.

Mott also became sufficiently intrigued with the prospects of advancing solid-state physics that he switched his main interest to it at just about the time, in 1935, that I had obtained a faculty position on the junior staff of the University of Rochester. There then began Mott's outpouring of wonderful articles and books, starting with one devoted to metals and alloys and coauthored with Harry Jones. It was soon followed by another with Ronald Gurney that extended the area of interest to ionic crystals. The flood of productivity carried on until the end of the decade.

Almost every item Mott produced in this period was a gem. Everything had to be read and reread. Otto Stern once said, with particular reference to Einstein, that some physicists have the gift of what he called the *Trüffelschwein*, being able to select problems with special insight. Mott displayed that gift in his work, not least in that of the 1930s.

General Electric Laboratory: Pittsburgh Summer School

In 1938, while at General Electric and with much freedom to develop the consolidating book on solid-state physics begun in Rochester, I was pleased to receive an invitation from E. Hutchisson, head of the Physics Department at the University of Pittsburgh, to join a lecture program at a summer school he was holding on the campus. The principal visitor-lecturer was to be Nevill Mott. I accepted with alacrity and had a wonderful two weeks with him and the group which joined him there.

Since I was somewhat freer of responsibilities to my home laboratory than most of the others, Mott and I found ourselves relatively alone over weekends and took many walks together through the verdant Pittsburgh hills. The great steel mills were, for the most part, silent at that time as a result of the continuing economic depression.

This was Mott's first visit to the United States. He was deeply impressed with some of the grosser aspects of the country, highly accentuated by the special 'stark beauty' one sees in some areas of industrialized Pittsburgh. One day while walking in the hills, we encountered a fluttering group of butterflies. With a laugh he said, 'Dammit! Even the butterflies are bigger than ours.'

Rochester and Schenectady

At the end of the session, Mott and I took trains, first to Rochester and then to Schenectady. Mott had just developed his basic theory of the photographic image and was giving a series of lectures on the subject. We were warmly welcomed at the Eastern Kodak Laboratories in Rochester, where there were at that time a large number of research chemists from England. I have always thought that Mott's contributions to the understanding of the photographic image represented one of the high points of his many contributions to the field of solid-state physics and deserved the Nobel prize alone, and much earlier in his career.

Mott's journey on that occasion ended in Schenectady, where Ruth joined him. He visited the General Electric Research Laboratories and was given another warm welcome.

World War II

Mott and I then began an on and off correspondence which continued throughout his lifetime. In the meantime, World War II intervened and both of us became deeply involved in our own ways in contributing to whatever problems were at hand. Regular correspondence with Mott was interrupted. Basic research was either intimately connected with applied work or had to be carried out in relatively rare spare moments.

European Assignment

In the late winter of 1945, the Secretary of Defense asked me to join a small team under the leadership of Howard P. Robertson, who had spent most of the war in London as part of a British–American intelligence team analyzing information received from the continent by various means. Robertson had agreed to establish a small intelligence office in Germany after the occupation began, in order to obtain deeper insight into German technical developments. As a result, I landed in Paris in late April of 1945 just as the war was ending. My orders permitted me to travel anywhere that I wished within British–American controlled domains by military vehicle. After getting settled in Paris and witnessing the noisy VE Day celebrations, I flew to London, reached Mott in Bristol by telephone and we were soon reunited there. For reasons which I had never clearly understood, despite all his remarkable abilities, he had never become quite as deeply involved in the frontier research and development programs as I had, although he had carried on some very important work in the field of ballistics and related areas. His colleague Herbert Skinner had played a major role in bringing silicon as a semiconductor back into use in radar in the form of a rectifying diode. Mott could have contributed much to that activity. In any event, the way to the future was open once again.

Postwar Period: Meeting at Göttingen

Early in 1949 I received an invitation from Robert Pohl, Head of the Physics Institute at the University of Göttingen, devoted to solid-state research, to participate in a one-week conference where matters related to crystalline defects, particularly color centers, would be reviewed. Mott was to be a principal participant. As a result, Robert Maurer, a lifetime colleague, and I traveled to Göttingen that summer by way of Amsterdam and Hamburg. The new stabilizing German currency had just been introduced, and the effects on trade were evident everywhere.

The activities at Göttingen were very lively and moving ahead as rapidly as circumstances permitted, with the encouragement of the British occupying authorities. The number of visitors to our conference from other places was small but the discussions were spirited and friendly. Mott had restored the Bristol Laboratory as a major international center for solid-state research and discussed much of the work going on there. Both Rudolf Hilsch and Walter Schottky participated as visitors from Erlangen. In the meantime, we were helped in every possible way by the laboratory staff, particularly Heinz Pick, who never having become a member of the National Socialist Party, had spent the war years as a factory technician and was now back in academic work. He would eventually become one of the major leaders in solid-state research as head of an institute at the University of Stuttgart.

It was a special pleasure to meet both Hilsch and Schottky, whom I had known very well through their publications. I had envisioned Schottky as a severe humorless individual, but he turned out to be exactly the opposite.

The conference extended over a Sunday, when the participants were free to do as they pleased. Mott, Maurer and I wandered through the university town, which was much in need of a facelift, since paint had not been available for many years. We took a long walk through the countryside, and on returning to the town, we found a moving picture theater showing one of the first postwar German films. It featured Hildegard Neff and dealt with the careers of a group of friends who had experienced the ups and downs of the previous two decades. Much was made, in a semihumorous way, of the tensions which developed within the German community itself as it experienced the difficult postwar years. Mott, who along with his fellow citizens in Britain had encountered much hardship during the war, seemed to relish this part of the film and chuckled frequently as it displayed many examples of human frailty and selfishness, as well as courage.

The British military still maintained a portion of its headquarters in Göttingen. The commanding officer invited Mott and me to a lunch. Otto Hahn was a frequent visitor to headquarters and appeared for a brief period both to greet us and discuss some personal matters with the commander.

The University of Illinois

My wife and I decided to move to the University of Illinois in 1949 as a result of receiving a magnificent offer to start a large solid-state physics group there. Since the move had been all but accomplished when I met Mott in Göttingen, I invited him to spend a period with us at his earliest convenience. As a result, he joined us for a month in 1950 and took the opportunity not only to explore in great depth every segment of the university which interested him, but to travel about the country where he was a most welcome guest at many laboratories. As expected, he continued to display that remarkable vision which permitted him to offer special insights into ongoing research programs, insights often beyond those carrying out the work. As also might be expected, he greatly enjoyed meeting again some of the postdoctoral fellows who had worked with him at Bristol and were spending a period of time at Urbana doing advanced research.

Solvay Meeting

In 1951 the organizers of the Solvay Conferences in Brussels, Belgium, decided to sponsor one in solid-state physics. The program was planned under the leadership of Lawrence Bragg and Nevill Mott, who became intimately involved. A relatively large international group assembled for a week, carrying on endless discussion of every aspect of solid-state physics of ongoing interest. We feasted like royalty on high Belgian cuisine and some of us discovered that we had more food allergies than we suspected. This trip made it possible for me to visit Bristol once again and provided an opportunity to bring a gift of the newly developed long-playing recordings of chamber music which Ruth and Nevill enjoyed so much. Ruth was, at that time, participating in the reconstruction and playing of antique musical instruments such as spinets and clavichords.

Mott Leaves Bristol

In 1954 Mott decided to accept an appointment at Cambridge University as head of one of the colleges and assume a new leadership role in British academic life. To commemorate the occasion, his colleagues held a scientific convocation in Bristol, inviting many individuals from home and abroad to celebrate his contributions to the university and to the advancement of science. This went beyond being merely a productive, festive occasion but, in a sense, represented one climactic point in Mott's complex scientific life.

Once he arrived in Cambridge, Mott became deeply involved in administrative work, taking responsibility for long-needed changes in a complicated academic environment. He obviously stayed closely associated with ongoing research and made his own contributions. But gone was the relatively unlimited freedom he had once had. We continued to exchange letters regarding minor matters of mutual interest. I took several occasions to visit him and Ruth while attending meetings in England.

NATO Period

My wife and I spent much of the years 1959 and 1960 in Paris, where I headed the Office of Scientific Research at NATO, an organization started under the stimulus provided by the launching of space satellites by the Russians in 1957. The appointment gave us an opportunity to visit London and Cambridge several times. Moreover, Mott and I were graciously given honorary degrees by the University of Reading in 1960, along with the customary academic pomp. Mott's aged father, who had been on the scientific faculty of a secondary school, attended the ceremonies. He had started research for an advanced degree in Cambridge under J. J. Thomson. When his experiment, which had been suggested by Thomson, led to a dead end, he was denied a second opportunity and shifted to an alternate academic life. His interest in events in Reading focused on his now very distinguished son, who was living the life that he had originally hoped to achieve.

Mott's Election to the National Academy of Sciences

During this period I took the opportunity to nominate Mott for foreign membership in our National Academy of Sciences. It was a pleasure to see that he was not only supported unanimously by the members of its physics community, but was promptly elected by the general membership, as I had hoped.

Later Years

In later years our mutual interests shifted mainly to biographical and historical topics as he began preparing his own biographical material. As a result, I did not follow closely the work which he carried out on non-crystalline solids, for which he was eventually awarded the Nobel prize. I became deeply involved, first as President of the National Academy of Sciences and then as president of a research university. Our face-to-face meetings were very few but always rewarding as a result of so many years of association.

Meetings at the University of Pavia

Our two most memorable meetings in the later period took place at the University of Pavia. Although the events involved much activity on the part of the administration and faculty of the university, it became evident that individuals from other institutions in Italy who had worked with us during their careers had played an important role in the planning. Principal among them were Fausto Fumi of the University of Genoa, my first postdoctoral fellow from abroad, Franco Bassani of the Scuola Normale Superiore in Pisa and Gianfranco Chiarotti of the University of Rome. Fumi had been intimately involved in the 1950s in the creation of the study center, Villa Monastero, on the eastern shore of Lake Como. It became a Mecca for solid-state and other physicists.

In 1977 the University of Pavia awarded honorary degrees to both Mott and me, as well as several others, in a very colorful ceremony. The situation was made somewhat complicated by the fact that university students throughout Italy were staging a revolt at that time. Chiarotti, who had been attempting to serve as ombudsman to the students in Rome, arrived looking pessimistic and exhausted.

The second meeting in Pavia occurred in 1987 and was devoted to a survey of various aspects of the foundations of solid-state physics, as viewed by the participants. The gathering was relatively large and the new students openly displayed no revolutionary inclinations. It was a very happy occasion.

Last Links

Our last period of communication emerged in a somewhat remarkable way. The leaders of a branch of the Vatican became interested in the growth of world population and its possible consequences. I assume that by this time

Mott had been elected to the Pontifical Academy and was regarded as one of its most highly distinguished members. He wondered if I could supply him with some literature dealing with the topic that would be relevant to the assignment.

Fortunately, I had immediate access to a great deal of material, including a newly published book on the subject by one of our faculty members at the Rockefeller University, Joel Cohen, who was able to offer predictions of patterns that might emerge in the future. All of this seemed to be moving along quite well. But then one day I turned a page of the *New York Times* to the obituary section and realized that all had ended. Nothing in our correspondence had indicated that this remarkable man was facing an immediate personal crisis.

PROFESSOR F. SEITZ is President Emeritus of the Rockefeller University and Past President of the National Academy of Sciences. He has had a long career in the field of solid-state physics as well as in academic and scientific administration. He and Eugene Wigner stimulated the field substantially in the early 1930s by developing a method for obtaining semiquantitative solutions of wave functions in crystals.

Mott Family Life

Alice Crampin

Nearly half a year has passed since my father's sudden death. Although he was full of years and honours, and in his last months never really very well, it still seems inappropriate for me to refer to him as if he were not coming back; so I will write a little about our family life as I remember it.

My father was the pivot of his family's world. This was not because he was a dominating authority figure or obviously a paterfamilias, but rather because he seemed to be at home and at ease in his milieu, pursuing his career and enjoying it. In Bristol, where my early childhood was spent, we lived across the way from the H. H. Wills Physics Laboratory, and so my father's, and indeed my mother's, home and public lives were closely intertwined. It was clear that they were established and respected figures in their community, and I think we children drew the inference that, therefore, we were established and respected as junior members of that community.

My father certainly allowed me to think I was someone special. I could easily wheedle my way into accompanying him on his striding walks across the hills around Bristol. We went to the Cotswolds, the Mendips, to Goblin and Monckton Combes and to the Upside Down Mountains – where were they? I always slowed him down, but he never complained. I could persuade him to let me play with the meticulous and intricate cardboard model of Chartres cathedral he had made years before. I broke it. I begged my parents to let me play with various other precious toys from their youth; all I damaged. I was never reproached. I did reproach myself and, years later, adopted a far more 'O, do be careful' tone with my own children. My parents appreciated their possessions deeply, but they felt they should be used. They accumulated things that interested and pleased them, but never seemed to grieve greatly over their loss.

Stuart House, in the Royal Fort, Bristol, where we lived throughout my early childhood, was full of elegant and well-loved objects, contrasting with equally many battered and casually maintained ones. Parts of the house never seemed to have recovered from wartime austerities. There was bleak, but hygenic, lino underfoot upstairs, and brown marks on the ceiling which were legacies of the leaks after the hard winter of 1947. Electric light-bulbs lacked shades, and carpets were never wall to wall. On the other hand, there were lots of books and many pictures, and many favourite plates hung on display. There were wonderful old tiles around the fireplaces downstairs. The rooms were filled with my mother's instruments; piano, clavichord and harpsichord. She

Nevill in his thirties.

had studied with Arnold Dolmetsch, and one of her passions was playing early music on keyboard and recorders with her family and friends. Another was her professional interest in the classics; she kept up her part-time teaching of Latin and Greek until she was well past eighty.

One cabinet in the corner of the drawing-room housed my father's collection of foreign dolls, brought back for the family from his many trips abroad. I was allowed to play with these on special occasions and, being very much a doll lover, I adored this, leaving them somewhat the worse for wear in my enthusiasm. I think the collection started with the dolls my parents brought back from a memorable prewar trip to Russia. The collection received a great boost in 1953, when my father went to Japan and came back with a beautiful big box full of carved wooden dolls and other souvenirs. Many of them were given away, but others were kept. I remember for the first time becoming intensely aware of Japan, and its scenery and culture. I was too young to have war memories and I saw it as the land that loved cherry trees, which were my favourites, too.

Stuart House dated back to the seventeenth century. Its cellar had a secret passage entrance; alas, it was thoroughly blocked up. In the Civil War, the Royal Fort had been held by Prince Rupert. An archway of that period still survived, together with other remains of fortifications, and some bombed-out houses that in the late forties and early fifties had not yet been cleared. Next to our house was Royal Fort House, used as the University's Education Department, and behind it lay extensive gardens, known to us as the Big Garden, where we children could roam freely.

Our house had an enormous kitchen, with an open fire and Welsh dresser, and numerous back kitchens and pantries leading off it. Here presided Peggy, our housekeeper, together with Bud, who worked long hours as a bricklayer, and their son Tony, our playmate. The kitchen was the place for cosy gatherings, of which there were a lot, and warmth and comfort, including mustard footbaths. And for the earache, from which I suffered a great deal as a child, hot onions applied to the ear by Bud.

In our earliest years at the Royal Fort my aunt Mary Horder (now Madame Friedel) also lived with us. She had her fretsaw set up in the back

H. H. Wills Laboratory, and Royal Fort House, Bristol.

reaches of the kitchen, where she worked on turning out the wooden toys she sold to the Heals shop in Tottenham Court Road. Her friends and colleagues in this enterprise, Maita Frank and Diana Murray Hill, were often with us, and even when Mary and Diana took a house down the hill at 2 Bedford Place, we still saw much of them. A memorable event in our childhood was a most elaborate production of *Toad of Toad Hall*, directed by Diana, in which the whole family took part. This was initiated in the first instance to provide amusement and stimulation for my sister, whose constant ill health in those years meant that she was frequently absent from school. Mary made the most wonderful costumes for this, including very realistic, if rather airless, animal masks.

The Stuart House drawing-room was big enough for the staging of this drama, and for all the constant stream of slightly more formal social life that flowed through the house. An amazing extending dining-table and curtains that could divide the room enabled us to cope with all eventualities. At Christmas the table, at its maximum extension, seemed easily to accommodate the whole clan and more. In the early 1950s my paternal grandparents retired to Bath and my father's sister Joan and her husband also lived there for a while, so there were often a lot of us about. Aunt Joan helped us to make up rhymes and verses, weaving our family life into amusing ballads.

Stuart House made a beautiful setting for social gatherings of any kind. Sociability and entertaining just seemed to happen in a relaxed way, and my parents thrived on it. Maita and Charles Frank, Cecil and Isabel Powell, Jack Mitchell, Jacques Friedel, my godmother Alice Terry, Liselotte Leschke, the Tyndalls, Pipers, Kittos, the Closs family, and many, many more all seemed to me to be constantly coming and going through our house. I have heard that hungry research students said to each other: 'Let's go over to Stuart House; they feed you there.' I do not know what was provided for them to eat, because these were the days of rationing; but Peggy, who hailed from Northern Ireland, would have had the situation well in hand. She could do miraculous things with the simplest ingredients. I remember, too, that at one point in the early postwar years, we were very surprised to receive an enormous parcel of delicacies from kindly Americans.

Cecil Powell was well known to us children as the man whose balloons came down in the middle of the night in far away places, occasioning exciting long-distance telephone calls in the small hours, and consequent delightful nocturnal disturbance. He was also the source of offcuts of balloon silk which could be pressed into use for highly effective angel costumes for nativity plays at Christmas.

An early visit in the postwar years, when toys were rather few and limited, was from Percy (E.A.) Andrade, who became legendary to us children by picking up a recorder-cleaning mop and investing it with magical properties as the 'gollyarchaphone'. Another memorable visit, because my father was so pleased to see him, and because he was so nice to children, was from George Gamow.

Nevill with George Gamow in Nevill's garden at Stuart House, Royal Fort, Bristol (1954).

My father was away a lot, and always busy, but when he was at home he used to read to us. He read Kipling to us and the works of E. Nesbit. I think these authors reminded him in different ways of his own childhood. He introduced us to H. G. Wells and Sherlock Holmes when we were slightly older, and Shakespeare and T. S. Eliot and Iris Murdoch – well, both parents were very much in favour of reading in general! I seem to remember that they reminisced about queueing at the bookshop for the latest T. S. Eliot as the volumes came out before the war.

My father had a whimsical fondness for elephants, which never left him. He always had a little brass elephant in his room. Among Kipling's *Just So Stories*, 'The Elephant's Child' was definitely his favourite. We knew we were the best beloveds, as in his later years his grandchildren also became. He told us elaborate made-up stories about elephants in the Lab. The Bristol Lab was a rather megalithic building, especially from a child's point of view. It seemed

Nevill in the garden of the Master's Lodge, Caius College (1963).

to have an affinity with elephants, and the stories flowed on and on naturally. The elephants lived in the lofty lecture theatres, one of which is, I believe, now called Mott. My father had an elephant joke which I thought highly sophisticated when I first learnt it, though now perhaps its charm is dated. An international symposium on the elephant is held, and the participants prepare papers: the Englishman contributes *Elephants I Have Shot*; the German learnedly discusses metaphysical problems in *The Elephant: Does It Exist?*; the Frenchman writes of *Les Amours de l'Elephant*; the Russian gloomily narrates *The Tragedy of the Elephant*; while the American leads them all forward with *Bigger and Better Elephants*. Perhaps there is still a point here, concerning naming and meaning. For some reason my father always claimed the only phrase in Russian he had managed to learn was 'the elephant eats bread'.

My father was always notorious in the family for his love of chocolate. One year I was given a doll modelled in chocolate. Chocolate enthusiast though I was, I could not eat it because it was a doll. I kept it in a box on a shelf hidden behind an armchair, and went to look at it from time to time. One day, the feet were missing; then the legs went. Later there was only the head. There was a definite suspect: however, I think I was quite glad to have the problem resolved for me.

I remember my great enthusiasm in 1954 at the prospect of moving to Cambridge. I am told I cheered up my parents, who had regrets about leaving their beloved Bristol, by my excitement on being taken to stay at the old Garden House Hotel with the river gurgling outside. 'We are going to Cambridge,' I remember boasting to my best friend at school. 'So are we,' she replied, taking the wind out of my sails. So we sat the entrance exam to the Perse Girls School together, and continued our friendship throughout our schooldays. My friend's father was taking up an appointment as Warden of Bassingbourn, one of Cambridgeshire's famous village colleges.

In Cambridge we lived in Sedley Taylor Road. The house was initially let to us by Sir John Cockcroft, for whom it had been built in the late 1920s, the architect being Mr H. C. Hughes. It had a very long road frontage, so long it appeared low in proportion to its length, and it was full of windows and light. It was built in yellow brick with red Norfolk pantiles. The extra length was because it had been substantially extended. This led to some peculiarities in the design. The original sitting-room had been bisected to provide access to an enormous airy but icy drawing-room, built on beyond it, so the original room was now awkwardly shaped, with a through draught and a fireplace in the corner. There were various other maddening features of this nature, and the house combined great charm with a certain amount of discomfort which just had to be put up with and lived around. It was, however, a house full of character, and the drawing-room was wonderful for entertaining, especially in summer. There was a half-acre of garden, which we all adored. Later on, when the Cockcrofts decided to sell the house, my parents bought it, only to find that they themselves were moving to Caius College to live in the Master's Lodge, while the Cockcrofts were returning to Cambridge to take up the Mastership of the new Churchill College, and needed a home until Churchill's Master's Lodge was built. My parents let the house to the Cockcrofts, before living in it again themselves for another fifteen years after they left Caius.

The Cockcrofts had planted many fine fruit trees when they first lived at Sedley Taylor Road, and to us fell the happy task of harvesting the fruit. It seemed to take all summer. There were about twenty plum trees; we started with the Early Rivers and went on through types unknown, but delicious, to the Victorias, the Czars and the damsons. My mother had three different chutney recipes and the cauldrons seemed constantly bubbling. Then there were the apples. I have never tasted such wonderful apples and I still do not

Nevill the sportsman (1975).

know what they were. The Friedels used to come to spend much of the summer with us, and when Jacques and my father were not pacing the lawn, talking the mysterious language of physics, they were picking fruit. Alas, the only problem was the wasps, which tormented us all and terrorised the young Friedels, causing the physics sessions to be rudely broken off.

Both my parents were enthusiastic gardeners. There was a rough division of labour: my mother planted and my father mowed and cut down. Everybody was supposed to weed. My father was an enthusiastic bonfire maker. It went back to his youthful bicycling and camping holidays when fires were lit to boil kettles. Sometimes he was too enthusiastic; an apple tree suffered. He also used to collect up litter on country walks and burn it; but not any more after an incident when a fire got mildly out of hand after he thought he had extinguished it. To his horror, someone called the fire brigade. Never again.

The country walks continued. There was an enchanting place called Nine Wells only a few fields away from suburban Sedley Taylor Road. Here water came gurgling out of a mossy bank in a copse to start a little stream. We explored the Gog Hills and Wandlebury; we walked along the Cam from Quy to Lode, and along Fleam Dyke and we explored the violet wood at Orwell. We still missed the West Country, and sometimes had half-term breaks in the Cotswolds. All the same, we became fonder and fonder of north Norfolk, and more half-term breaks were spent at Blakeney and Morston, then at Burnham Market. This culminated in the purchase in 1964 of a tiny cottage at Burnham Market, which was much loved, much lent and much used.

During my father's years at Caius he took a great interest in the East Anglian settlements where the college held livings and in the incumbents of those livings and their families. I particularly remember getting to know the lovely town of Lavenham in this way and I recall visits with my parents to Canon and Mrs Cotton there in their beautiful old rectory, which on a fine summer day in the early sixties provided the perfect setting for a church garden party.

Nevill with family and friends outside his house in Aspley Guise on the occasion of his eightieth birthday (1985).

The lodge at Caius also provided a fine setting for the many gatherings that my parents considered it both their duty and their pleasure to host. People liked being asked to parties at Master's lodges and they were indeed rather grand; with the wonderful Caius staff headed by Lionel Rumbelow, the butler, always so kind and so dignified, bearing in the specialities of the college kitchen, such as *crème brûlée*, on silver salvers. Now in my late teens, I tried to learn the gracious arts of circulating at sherry parties and talking in the proper turns to one's neighbours at dinner parties. To me, my parents seemed naturally good at this, though both told me of their shyness in youth, and I believe my mother was beginning to be nagged by an insidious deafness.

Caius Lodge was enormous and labyrinthine, and composed of parts built at many different periods. Bits have now been taken away and incorporated into the college, a loss to any future Master's children. There were two magnificent reception rooms in the eighteenth-century part of the house, full of college furniture and portraits. There was a splendid room in the sixteenth-century wing, called the John Caius Room, with lovely oak panelling. Then, there were the five bathrooms, some with amazing antique fittings. There were various pantries that seemed to lead away into the bowels of the college and from whence not a few cockroach beetles emerged, to my great interest. I was doing A level biology and we dissected cockroaches. There were secret places where you could get onto the roof. I don't think it was encouraged, but it was quite safe. You could look out on Caius Court, Gonville Court, Trinity Lane or the beautiful enclosed garden of the lodge, gardened by Mr Muggleton, with its ancient wisteria on the house and the original Gate of Humility forming part of the end wall. There was also a magnificent ailanthus tree. Unfortunately, in high summer it gave out a stuffy smell. Presumably our forerunners in the lodge had known that it was then time to retreat to the other lodge owned by the college, at Heacham by the sea; but by our time this exodus, once compulsory for Masters of Caius, had long since lapsed.

In the lodge, at times of carousing in the college hall, one could hear coming from overhead the 'song of them that triumph, the shout of them that feast'. In the lodge below, we women were eating our frugal suppers, for those were the bad old unintegrated days.

My parents took a great interest in the college chapel and its services. My mother devoted much effort to doing the chapel flowers, and they both loved the choral evensong there. So did I. They were enthusiastic supporters of all the musical life of the college, and very good it was too. Martin Neary, now at Westminster Abbey, was the organ scholar of the time.

Caius College played host to the Pugwash conference that my father helped organise during his time there. Both my parents threw themselves into this, and socially at least you could feel it was a great success. My mother organised the ladies' programme, but it was Marks & Spencer that our visitors from Russia really wanted to see, and see again. At Caius we were very strategically placed for the central Cambridge shops.

I set out in these notes to describe my father in the domestic landscapes

Nevill in full flow at his house in Aspley Guise (1989).

where he placed himself, and in the event seem to have been describing one long social whirl. It was not always quite like that; but my parents loved to promote conviviality and thought it their duty as senior members of a university community to do so. In Bristol, their location so close to the physics lab, together with the dynamism of a young department, facilitated community life. I believe they adored that. In Cambridge it took more effort to make things happen. My mother's work with the Cavendish Wives Club, which she founded, is a case in point. She had felt herself what it was like to be the potentially lonely young wife of a graduate student; she thought this was an avoidable evil that she could do something about, and my father supported her fully in the enterprise.

I took my father for a visit to Burnham Market shortly before he died. Though he could hardly walk, he was so delighted to find himself in Norfolk again. We attended an open-air performance of *A Winter's Tale* in the garden of Westgate Hall. 'My dear people,' my father said to my daughter and me, 'I am so happy to be here with you.' He looked around hopefully to see if he could see any of his old acquaintances. None was in sight, so he settled down to making new ones, as he sat expectantly in the audience.

My mother and father were naturally open-hearted, always well disposed to new people they met, and extremely loyal to their friends. Though they could be subtle, 'treasons, stratagems and spoils' were alien to them. Their straightforward approach to their objectives led them to try to be community builders wherever they were involved. Perhaps this book, so kindly put in hand by Ted, is testimony to that.

ALICE CRAMPIN, younger daughter of Nevill and Ruth, was born in 1943 and educated at the Perse Girls School and New Hall, Cambridge. She worked in applied economics research before raising a family and is now involved in several fields of voluntary work in Bedfordshire. She is married to Michael, an Open University Professor of Mathematics, and is mother of Nevill's three grandchildren, Emma, Edmund and Cecily.

Nuclear Arms and Pugwash

Preventing Nuclear War

Joseph Rotblat

Nevill Mott was not only a great scientist, he was also a man who cared for humanity. He was profoundly concerned about the impact of science on society, particularly after the development of nuclear weapons, and was very active in the movements of scientists aiming at averting the danger of a nuclear war.

Most of the scientists on the Manhattan Project who were consulted on the issue were against the use of the bomb on a populated city. Having failed in this, they mounted a campaign to prevent further such acts. Fearful of the consequences of a nuclear arms race which they foresaw, the scientists (mostly those who worked on the atom bomb during the war years) set up organizations, the Federation of American Scientists (FAS) in the United States (initially the Federation of Atomic Scientists), and the Atomic Scientists Association (ASA) in the United Kingdom, with the main objective of ensuring international control on nuclear energy in both its military and peaceful uses.

Nevill Mott was a member of the wartime Policy Committee which supervised the atomic bomb project. Though not directly involved in research on the bomb, he realized the potential danger of a nuclear arms race, and he joined the ASA when it was formed in March 1946. The aims of the ASA were to bring to the notice of the public the true facts about atomic energy; to make proposals about the international control of atomic energy; and to help shape the policy of the United Kingdom on all matters relating to atomic energy.

Nevill Mott was the first president of the ASA; he retired from this post after two years, but returned for another term as president in 1950–51. Throughout the existence of the ASA he served as an honorary vice-president, but unlike most of the other vice-presidents he took an active part in the work of the ASA, both in its organization and in the substantive programme of activities, attending meetings, debating the issues and contributing articles to the ASA journal.

The list of honorary vice-presidents included the most eminent scientists in the country, as well as government advisers and leaders of the nuclear energy authority. These celebrities were invited to give respectability to the ASA, but this turned out to be a millstone around its neck. Some of the vice-presidents were strong proponents of nuclear weapons and they stifled any criticism of the British government when it decided to make these weapons. It made the ASA rather impotent and eventually it was decided to dissolve it. By then, in 1959, the Pugwash movement was in full swing, and most of the ASA activities were taken over by its British group.

The Pugwash Conferences on Science and World Affairs began with the proclamation of the Russell–Einstein Manifesto in July 1955. It contained a strong warning that a nuclear war could mean the end of the human race, and it called on scientists to assemble in conference to find the means to prevent such a catastrophe. In earlier discussions between the FAS and ASA it was recognized that, in order to have an effect, scientists from both sides of the Iron Curtain would have to be involved, but this was impossible during the Stalin regime. The Russell–Einstein Manifesto, initiated by Bertrand Russell, came just at the right time.

The conference called for in the manifesto was held in the Nova Scotian village, Pugwash, in July 1957. It was a great success, and the participants decided to continue with such meetings, using for their title the name of the place where the first meeting was held. By 1997, 230 Pugwash meetings had been held all over the world. The important achievements of Pugwash in preventing a nuclear war and halting the nuclear arms race were recognized by the award of the Nobel peace prize in 1995.

Nevill Mott was very active in Pugwash both on the international arena and in the British group. The running of Pugwash conferences was in the hands of the Continuing Committee (later renamed the Pugwash Council) and Mott served on it from 1959 to 1963. In 1962, two international Pugwash conferences were held in the United Kingdom, in Cambridge and in London. Nevill presided over both of them. In the earlier years, the UK government (like the US administration) was suspicious of Pugwash, viewing it as a communist front organization. But when the government became convinced that it was genuine and, moreover, an important channel of communication between East and West, it attempted to influence the discussion and the selection of participants. Nevill Mott was successful in his effort to maintain a balance: financial support from the government but complete independence of Pugwash.

Nevill hosted the Cambridge conference as Master of Caius, where the meetings were held. In London he chaired the historic opening session, which had on its platform many of the signatories of the Russell–Einstein Manifesto, including Bertrand Russell himself. The conference was formally opened by Lord Hailsham, then Minister for Science, who also read a message of greetings from Prime Minister Harold Macmillan. Russell was by that time deeply involved in antinuclear activities, as chairman of the Committee of One Hundred, and had been arrested for the sitting-down demonstration on the steps of the Ministry of Defence. He feared that he would get a cool reception from the 'respectable' Pugwash, but instead he got a long standing ovation, which in Russell's recollection included all the participants, save Lord Hailsham: 'Weighed down by office, he sat tight.'

Nevill Mott participated in a number of Pugwash meetings, the last one was in 1985 on Reagan's Star Wars project. He maintained an active interest in Pugwash to the end, as a trustee of the British Pugwash Trust. His last public intervention for Pugwash was to refute false allegations made in a British newspaper, after the award of the Nobel peace prize in 1995.

PROFESSOR J. ROTBLAT, FRS, is Emeritus Professor of Physics at the University of London. He was the joint recipient, with the Pugwash Conferences on Science and World Affairs, of the 1995 Nobel peace prize. He was President of Pugwash until August 1997.

The Oxford Research Group

Scilla Elworthy

Professor Sir Nevill Mott was on the Council of Advisers of the Oxford Research Group (ORG) from 1984 until his death. Fellow members of the council included Professor Frank Barnaby, former director of the Stockholm International Peace Research Institute; General Sir Hugh Beach, former Master General of the Ordnance; Lord Habgood, former Archbishop of York; Mary Midgeley, lecturer in philosophy at the Universities of Reading and Newcastle; and Ben Whitaker, former director of the Minority Rights Group.

The Oxford Research Group is an independent organisation funded by charitable trusts to carry out research into nuclear weapons decision making, and to bring decision makers from all the nuclear nations to meet with their critics and to discuss disarmament.

Sir Nevill, an outstanding physicist who infinitely regretted the military use of great scientific discoveries, was very interested in the Oxford Research Group's approach to the dangers and challenges of the nuclear arms race. The Group's approach was first to discover the processes of decision making behind the nuclear arms race, and to identify the individuals involved (information previously shrouded in secrecy); then to initiate informed, non-confrontational dialogue between critics of their policies and the decision makers themselves. Nevill personally took part in ORG's Dialogue Project, as it was called, in the 1980s when over sixty groups throughout the United Kingdom were put in touch with, corresponded with and in some cases met a nuclear weapons decision maker, and his counterpart in China. The dialogue approach has since been adopted in America, and recently on a worldwide basis by the International Physicians for the Prevention of Nuclear War (IPPNW) in their Abolition 2000 campaign.

The Oxford Research Group will always be grateful for Sir Nevill's early support, which did so much to establish its reputation, and for his continued active interest; he always kept in touch with handwritten letters until shortly before his death.

DR S. ELWORTHY completed her secondary education as Herts County Scholar at Berkhamstead in 1961 and graduated from Trinity College, Dublin, in social science in 1965. After a Research Fellowship at the University of Bradford, she obtained a PhD on British nuclear weapons policy in 1993. She founded the Oxford Research Group in 1982 and holds central responsibility for its work, which includes leading delegations to capitals of nuclear weapons nations, including China.

Early Cambridge

Memories of the Professor's Secretary

Shirley Fieldhouse

By the summer of 1959 I had already been in Cambridge for six months in a temporary office job while looking for a personal secretarial position. I had come from the North of England seeking a change but unclear what it would be, except that it was likely to be personal secretarial work for which I had been trained. My only contact was my brother Martin, who had embarked on a postgraduate course at the Computer Laboratory. We both attended the Unitarian Church. There, Chubby (F. J. M.) Stratton (Emeritus Professor of Astrophysics and Fellow of Gonville and Caius College) was chairman. He informed the minister that the Master of Caius was looking for a secretary and thus the minister told me. I applied, but was unsuccessful. Some weeks later I had a letter; the Cavendish Professor of Physics needed a secretary. Would I like to apply? Mysterious! Unused to the ways of colleges and university departments and the fact that an academic could hold two positions, I enquired of my brother whether I should follow this up. Of course, was the immediate response. I applied and was invited to attend an interview at the old Cavendish Laboratory at 6 P.M. After meeting the Departmental Secretary, I was introduced to a big gentleman, tall and rather thin. The interview was quite brief. Eight days later I went for a second interview and was offered the job; great jubilation! Shortly afterwards, at the beginning of July, I reported for duty, thinking that the job would suit me for three years, little knowing that I would have the privilege of knowing the Professor for the next thirty-seven years and still be working in the laboratory at the time of his death.

The Professor's office was on the second floor of the Austin Wing, a 1939 block in the centre of the New Museums Site, behind the original Cavendish Laboratory which fronts Free School Lane. The office was above the hut, where Dr Max Perutz worked, and faced the Department of Metallurgy and the Computer Laboratory, which lay along Corn Exchange Street. At right angles to the Austin Wing, and adjoining it, was the High Tension Laboratory. Nearby was the Philosophical Library. The Professor had a large rectangular office. The two short walls and the wall opposite the windows were lined with bookshelves on which he kept a large collection of scientific and general books and current journals, among them the *Philosophical Magazine*. The comfortably furnished room held a large desk, a round table and two easy chairs. My office next door was lined with cupboards to a height of about 45 inches, and their broad tops formed a useful shelf around the office. The Imperial manual typewriter was screwed to a recessed shelf on the metal desk

The old Cavendish Laboratory in Free School Lane.

and could be folded down to provide a flat working surface. A few metal filing cabinets and a telephone completed the furnishings. Occasionally another secretary, with her own desk, shared the office. The telephone connected to the departmental switchboard, controlled by Miss Mary Frost in a cubbyhole in the dim entrance foyer. Whether one wanted an outside line or another extension in the same building, one had to ask Mary. Out in the corridor, on the opposite side to my office, was an alcove where the small departmental museum was housed. Next to that were the small offices for the departmental secretary and his assistant. To my right was the kitchen and the tearoom, used by staff and assistants alike, though sometimes the Professor had coffee served in his office. At the end of the corridor was the Rayleigh Library. The remainder of the building was made up of laboratories, workshops and offices. The department also occupied parts of the buildings along Free School Lane, including the basements which, as academic premises were at that time I think exempt from factory safety regulations, escaped the attentions of the inspectors who would surely have been horrified at the cramped conditions (and where were the fire escapes?) in which staff and students worked. The department also occupied the High Tension Laboratory, the Mond Laboratory for low temperature research and later the Phoenix Building.

Up to 1971, when the Professor retired from the Cavendish Chair, we endured a succession of building operations. First came the conversion of the High Tension Laboratory into the Cockcroft Lecture Theatre and teaching laboratories. The floor was of specially reinforced concrete – it all had to be broken up. The building, although built before the Austin Wing, was connected to it, so the vibration and noise penetrated our offices – to such an extent that for several weeks I sought refuge in an office in the Free School Lane building. Next came the addition of a fourth floor to the Austin Wing; that too created noise over many weeks. Finally came the demolition of the Metallurgy and Computer Laboratory buildings to make way for their new premises. The buildings were literally knocked down by the use of a ball which was swung against the walls. It was a fascinating sight. Equally memorable was the sound of cascading glass as the windows were demolished. Throughout these disturbances the hut which housed some of the MRC molecular biology research and then the departmental duplicating-room, continued to stand, as it does today.

My duties included taking correspondence in shorthand and typing scientific papers. Strong fingers and accuracy were essential. Erasure was difficult as the papers were typed double line spaced on to quarto sheets with three carbon copies. If we wanted more copies, then we used Gestetner wax stencils and gave them to Mrs Fi (Freda) Oakes to duplicate on her single duplicator, which had to cope with work from all over the department and the lecture handouts. The department was smaller in number at that time and I do not think the lecturers prepared such ample lecture handouts as they do now. The thermofax was the only means of copying documents and was used sparingly as it was time-consuming and the individual sheets had to be hung up to dry.

In the course of the next few years there were some changes in office machinery. The IBM electric typewriter appeared and I remember the advertisements for secretaries at that time, as an inducement, would boast 'IBM electric typewriter available'. The telephone system became automated to the extent that one could telephone other extensions on the site without going through the departmental switchboard. A second Gestetner duplicator was purchased! And the Xerox copying machine was installed.

Correspondence on departmental administration must have been done in the Departmental Secretary's office, for I do not remember my work being concerned with that. But there was a vast amount to do with scientific matters and with the Professor's outside interests: Taylor & Francis, Nuffield Physics, Pugwash conferences, Kurt Hahn and his associations with Atlantic College in South Wales and the Trevelyan Fellowships. Besides being editor of the *Philosophical Magazine*, published by Taylor & Francis, the Professor was much involved with the interests of the publishing company.

We worked from 9 A.M. to 6 P.M. with one and a quarter hours for lunch and on most Saturday mornings. The laboratory closed at Christmas until after New Year, for a week at Easter, the last week in June, and three weeks in August. The Professor did not often come in on Saturday mornings. Instead I went to his office in the Master's Lodge for dictation. He was a constant worker. A bout of influenza which confined him to bed in the lodge did not prevent him from summoning me to take instructions.

During the later 1960s the Professor collaborated with Dr E. A. (Ted) Davis on the book *Electronic Processes in Non-Crystalline Materials*. I was used to typing pages of references from manuscript, but on this occasion I decided the project was so important that it deserved to have all the references checked back to their source to ensure their accuracy. Thus I spent many hours among the Professor's journals, the Rayleigh Library and at the Philosophical Library. There in my pursuit of the original journals I found myself venturing into the basement amidst the arches which were evidence of the monastic buildings on which the Philosophical Library had been built. I felt my diligence was justified when I found (only a few) errors which had presumably been repeated throughout numerous scientific papers.

One of my duties was to assist Mrs Ruth Mott in running the Cavendish Wives Club, issuing invitations to the evening meetings at the laboratory to the wives of staff and academic visitors and attending the talks which a small committee of wives had arranged. These meetings were attended by Mrs Joan Fitch, the Professor's sister, who herself gave a talk on the National Trust's Neptune Appeal to buy coastline. Thus began my closer social connection with the Mott family.

In September 1961 I was looking for fresh rented accommodation. In order to allow me to look around at leisure, the Professor and Mrs Mott kindly allowed me to use the housekeeper's flat at the Master's Lodge, a room and kitchenette, in the attic of a building adjoining the lodge and overlooking the Master's Garden; it was a comfortable home for the next four months. Access was gained via the lodge's backdoor. One evening on my way up to the flat I happened to meet the Professor standing on a half-landing splendidly attired in a black silk gown about to attend a College Feast. On another occasion I met him in the kitchen, in shirtsleeves, helping his wife with the washing-up after a private function. I was fortunate to be included in occasional lunch parties in the Junior and Senior Parlours and also in the elegant Master's Dining-room. In the late 1970s the Motts generously lent me their holiday cottage at Burnham Market, Norfolk, where a friend, her children and I had a happy holiday, as did my parents another time. During the late 1960s, as a consequence of attending a fundraising event hosted by Joan Fitch, who was treasurer of the county branch of the Council for the Protection of Rural England (CPRE), I became a member of that organisation and after a few years found myself on the committee, then secretary from 1981 for nearly fifteen years. The Motts were members and their daughter Alice became a very active member of the Bedfordshire branch. In March 1996 the Professor supported a CPRE coffee morning in the Senior Parlour at Caius. The Professor knew of my active interest in another organisation, Amnesty International, and it was characteristic of him that he should remember this later in one of the Christmas cards which he and Lady Mott sent me. On this card was a colour photograph which he had taken of a window in All Saints, North Street, York, to which he had given the legend, 'I was in prison, and ye came unto me' (Matt. 25: 36). Later, when he was asked to send Amnesty Urgent Action messages, it was natural that he should ask me to type the letters.

In 1971 he retired from the Cavendish Chair and lost my secretarial services. He scrupulously avoided asking me for secretarial help for some months. But when it became evident that I had some spare time, I did some typing for him and this continued when we moved in 1974 to the new Cavendish Laboratory in Madingley Road and up until 1979, when I became secretary to the Departmental Secretary. From then on, as he had an office in the Mott Building, other secretaries in the building looked after him until he made the final move to the IRC in Superconductivity, where the secretary there took over. However, I continued to encounter him on his frequent visits to the laboratory and he was always happy to stop and exchange a few words. The lunch given to him by his colleagues Professors Yao Liang and Mike Pepper, in the Senior Parlour at Caius on his ninetieth birthday, was a very pleasant occasion. I was honoured to be sitting next to him and enjoyed his conversation and his witty response to Sir Alan Cottrell's toast. His ability to command attention never faded. Only a couple of years before his death, I was showing some visitors the laboratory museum; they were attending a conference at one of the colleges and were full of awe and admiration for the way in which he had chaired one of the meetings. We who knew him well enjoyed participating in all the causes for celebration: dinner at Sidney Sussex College in honour in his knighthood (1962), dinner at Caius for his seventieth birthday (1975), opening a magnum of champagne in the Mott Building on the award of the Noble prize (1977), ninetieth birthday lunch (1995) and the tea party for the launch of his book *65 Years in Physics* (1995).

Nevill Mott: Reminiscences and Appreciations

SHIRLEY FIELDHOUSE joined the Cavendish Laboratory in 1959 as secretary to the Head of the Department, thinking it would be a nice job for the next three years. She is still there 37 years later! Secretary to Nevill Mott until his retirement in 1971 and then to his successor until 1979, she then moved a couple of doors along the corridor to her present position as secretary to the Departmental Secretary.

Magister

Ian Nicol

Nevill Mott was not my professor in the science sense but when, as Cavendish Professor, he appointed me Secretary to the Department of Physics, he changed my career for the better and profoundly influenced it for many years. Whether or not he deliberately cultivated the image of a preoccupied professor, I have never been quite sure, but it was an image well wide of the mark. His powers of leadership extended far beyond his physics. He had all the right qualities. He thought strategically, established and stuck to his priorities, picked his staff, pointed the way and left them to get on with it; but was ready nevertheless with advice if asked, which together with his warm and generous spirit inspired great loyalty. I recall one example. Robin Day had arranged an interview with Nevill and arrived early, immaculately dressed and clearly expecting an immediate audience. A few minutes later the Professor's secretary asked me to entertain Mr (now Sir) Robin Day while the Professor finished his meeting. I tried to interest the visitor in the museum exhibits in the corridor outside the Professor's room but only engaged half his attention, the other half was fixed on the Professor's door. Thankfully the researcher left a few minutes later and Robin Day was ushered in. It was obvious that BBC stars were unaccustomed to being kept waiting, but for Nevill the research problem had priority. Later I asked Nevill how he got on. With that gentle, enigmatic smile he said he thought he had not helped Robin Day very much.

The forgetful image has given rise to several (I suspect apocryphal) stories about visits to conferences, and so on. Two episodes, however, come to mind. He drove Wendy, my wife, and myself to Hemel Hempstead to open a new school. The headmaster had been one of his students in Bristol and asked that the school could be the Cavendish School and include university-appointed representatives on its Board of Governors. Nevill supported the request and nominated me as the representative. It was founded as a coeducational technical grammar school – an experimental development in the new town. In the summer of 1962 it was ready to be officially opened by Professor Sir Nevill Mott and so we set off. His car was very small, an Austin 7 I think. Nevill was very big and crouched over the wheel, peering intently ahead. He drove with panache and not a little skill as we went along the A1 at a spanking pace, overtaking much larger cars with ease. His gear changing was unique and involved a great sweep of the arm which Sir John Barbirolli would have admired. Nevill must have gone on to a London meeting, as Wendy and I travelled back to Cambridge by train! On another occasion Ruth phoned me at home early one morning to say that their car had been stolen from their garage in Caius and Nevill had gone to London by train. To my relief, we found the car had been garaged in London overnight and Nevill had returned the previous night by train. Of course, the laboratory made a good story of this, but I believe the truth is that Nevill was having a tiring time and thought it safer to come back by train, but in his haste he forgot to tell Ruth.

Nevill was not enamoured of all aspects of Cambridge life, particularly many aspects of both university and college policies and administrative arrangements. In his view they were too insular and too heavily based on tradition. He frequently compared them unfavourably with his experiences in Bristol. He was particularly happy there and his move to Cambridge had been a difficult decision. As time passed, our talks ranged more widely, Faculty Board business, Nuffield science project, his work as UGC assessor for physics departments in some of the new universities founded in the 1960s, and the outrageous equipment lists which landed on his desk at a time when established and successful departments were being cut back to compensate, and much else. Two interests, however, occupied many hours of discussion. The political attitude to universities was changing; they were being asked to be 'more socially conscious,' academic science to be more closely related to the needs of industry and the national economy – a theme which is now rising again from the ashes of Wilson's white-hot technological revolution! Nevill was quick to see the real possibility of a more dirigiste approach to the funding of universities by Ministers and their civil servants at the Department of Education and Science (DES) and the University Grants Committee (UGC), and thought much about how Cambridge should position itself in response. An undoubted need was to speed up the administrative response time to changes in funding policy, and in this he was powerfully supported by his favourite lieutenant, Brian Pippard.

As Cavendish Professor and Head of Department, Nevill was a member of an institution, the Council of the School of the Physical Sciences (CSPS), which was established in 1927 to petition the university for a chair of theoretical physics. For many years the council met about once a term, mainly to propose further professorships. In 1962, however, it suggested that it should prepare the quinquennial statement for the departments within its scope. The request was turned down. Nevill perceived in this body, if suitably reconstituted to include the heads of all the departments and served by a full-time secretariat, a means of speeding up decisions. He pursued the idea with vigour and persistence and, having gained the support of the other heads of departments, the university authorities reversed their previous decision. Nevill persuaded the CSPS to accept an agreed set of priorities, which of course involved disappointments, and a responsible statement was sent to the General Board of the Faculties, where Brian ensured it was by and large included in the universities' submission to the UGC. The success of this more direct form of administration is shown by the fact that there are now four such councils covering all the faculties, each with its own permanent secretary.

At about this time the crisis of lack of accommodation on the New Museum Site led to proposals to put up towers onsite, a solution which horrified the experimental sciences and Nevill asked Brian to think about alternative solutions. Meanwhile I was becoming increasingly involved in developing the work of the CSPS, and Nevill appointed John Payne to assist me. He rapidly became involved with Brian in designing the new Cavendish Laboratory. But that is a story others may take up.

Which brings me to Nevill's other concern. How to ensure that some of his better research students could gain experience in industrial and government research establishments. Cambridge was badly placed geographically for the type of joint research programme suited to the training of a research student. Typically a student would disappear for about one-half to three-quarters of his three years, and was therefore largely cut off from seminar and postgraduate teaching programmes. By now I had become used to being invited to coffee or lunch in the lodge as a prelude to discussing a problem of concern to Nevill. He could agonise over a decision, particularly if it involved an individual. His kindness and consideration for the person had to be balanced against the wider university and college interest, and I suppose he just found it useful to check out the arguments with some neutral listener. If it was lunch then I knew I would be eating chicken Maryland followed by figs. Whatever criticism Nevill may have had of college affairs, it did not extend to the chef's skills with chicken.

By 1967 these discussions were increasing in frequency and the topic was mostly Cambridge science and industry. A number of events had come together since 1963 or 1964 which persuaded us the time had arrived for action. IBM wanted to establish their European Research Laboratories in Cambridge but the only site the county planners were prepared to accept was an unattractive one at Sawston. IBM went to Winchester instead. Though certain university departments were discreetly pleased, Nevill was disappointed. The Cole brothers had established a small company, called Metals Research, which worked closely with the Department of Metallurgy and, being successful, wanted to expand. They too were told by the planners to move out of the city, so the links with the department were weakened and eventually all but faded out. Word went out that Cambridge was unfriendly. This, against the background of government support for closer industrial–academic interaction! Nevill agreed to chair a small committee of scientists and engineers in the School of Physical Sciences with the task of finding solutions to the problem and involving industry, the city and county in its enquiries.

There were only three categories of industry for planners at that time, heavy, light and nuisance. As far as Cambridge was concerned, *light* was defined as employing no more than five persons. It was also planning policy that only research and development should be allowed to expand in Cambridge and that all industrial development should take place elsewhere in East Anglia. Nevill and I spent hours in Shire Hall explaining to the county planner and a special subcommittee of the County Planning Committee that we were not proposing a second Cowley but describing what would now be called a high-technology company. Most members of the subcommittee seemed to be country people, to whom industry was smokestacks and chemical plants. However, they did come round in the end. The city proved more understanding, and we had much support from local industry. By 1969 the point had been won and the university accepted the committee's report. Without Nevill's prestige and the distinguished membership of his committee, I believe it is questionable whether the university would have accepted such changes to long-established policies. Trinity College picked up the idea and the rest of the story is, as they say, history. Nevill took great pleasure in this success and I believe he considered the time and thought he gave to it fully justified.

Outside of his science, Nevill taught me much by example and by involvement. He was a joy to work for and a good friend. Most of us have a few people who cause the heart to sing a little when we meet them. Nevill had this effect on me.

DR A. D. I. NICOL obtained his PhD in the Crystallography Laboratory of the Cavendish Laboratory and was appointed Secretary to the Department of Physics in 1960, a post he held for six years. Subsequent appointments at the University of Cambridge were Secretary to the School of Physical Sciences (1966–1972) and Secretary General of the Faculties (1972–1984). He is now Emeritus Secretary General of the Faculties and a Life Fellow of Fitzwilliam College.

How he Saved the Laboratory of Molecular Biology

Max Perutz

In the late 1940s and early 1950s, with support from W. L. Bragg and the Medical Research Council (MRC), John Kendrew, Francis Crick and I built up a research unit for molecular biology in the Cavendish Laboratory. In 1953 this unit scored its first successes with Crick and Watson's solution of the structure of DNA and my demonstration that the phase problem in protein crystallography could be solved by the method of isomorphous replacement with heavy atoms. In the autumn of that year, Bragg retired as Cavendish Professor and moved to the Royal Institution in London. Nevill Mott was appointed to succeed him. When he found the laboratory too packed with people to bring along any of his own collaborators from Bristol, he found there one group with only tenuous connection to physics. That was ours.

Mott therefore wrote to the General Board of the Faculties asking them to serve us notice to quit. At the time, Bragg tried to draw us to the Royal Institution, but we did not think that it would provide the best environment for the development of our subject, and none of my colleagues wanted to move to London. Desperately alarmed at this threat to the continued existence of our unit, I wrote to the General Board that our work had entered a most promising stage and pleaded with them to let us stay at the Cavendish. To my immense relief, they responded by asking Mott to keep us there until alternative accommodation could be found. When Mott arrived in October 1954 he soon shared Bragg's enthusiasm for our work, and never again did he raise the question of our leaving the Cavendish.

Egon Orowan, a Hungarian solid-state physicist, had joined the Cavendish shortly before the war. Afterwards he founded a metal physics unit which was accommodated in a prefabricated hut in front of the Austin Wing. In the mid 1950s Orowan moved to MIT and his unit closed down. By then, our unit was expanding fast and we were becoming desperately short of space. I therefore called on the Secretary General of the Faculties and asked if I could inherit Orowan's hut. His firm answer was no, because, being a temporary structure, it was scheduled to be demolished. Today, over forty years later, the hut still stands. Mott helped me to get the demolition order overruled and the MRC refurbished it for us. The first two protein structures which earned Kendrew and me the Nobel prize were solved there, but our unit grew fast and the hut soon became overcrowded. I had to go round the department heads asking for a little bench space here and there. Crick did the experiments that led him and Brenner to the discovery of the triplet nature of the genetic code in a corner of the butterfly museum of the Zoology Department.

The stage seemed set to ask the MRC to build us our own laboratory. The MRC approved our plan enthusiastically, but it also needed the university's approval, and this proved far more difficult. Its administrators were strongly opposed to the establishment of non-university laboratories at Cambridge, because they would compete with the university for staff. The professor of biochemistry opposed our plans for motives that were all too obvious and wrote long memoranda to the General Board, prophesying that our laboratory, divorced from university teaching, would soon fossilise.

At that stage I began to realise that Mott was not only a brilliant physicist, but also an astute politician. He got himself elected to the General Board. Otherwise, he said to me, it was impossible to get anything done. Faced with opposition to our plans, he asked the board to have them examined by a subcommittee which would report back to the board. As its members, he got the board to appoint himself, Alex Todd, the forceful head of the Chemistry Department who also supported us, and the professor of biochemistry. Outvoted by these two powerful characters, the professor of biochemistry reluctantly signed a report recommending approval of our plans, and the General Board had no alternative but to follow.

But for Mott's effective intervention, our laboratory would not have come into being and we would soon have lost our world lead in molecular biology. I am grateful to have had this opportunity to put our debt on record.

DR M. F. PERUTZ, OM, FRS, came to the Cavendish Laboratory as a graduate student from Vienna in 1936 and remained there until 1962, when he became Chairman of the newly founded Medical Research Council Laboratory of Molecular Biology, where he still works. He and John Kendrew jointly received the Nobel prize for chemistry in 1962 for their solutions to the first protein structures, haemoglobin and myoglobin.

A Student's View

Maurice Rice

My first contact with Sir Nevill came when I was an undergraduate at University College, Dublin, in 1960 looking around for a place to start graduate studies in physics. I wrote to enquire at the Cavendish Laboratory and was overwhelmed to receive a very gracious and detailed reply from Sir Nevill in person, listing all the possibilities, from radio astronomy to particle physics. However, I had heard in Dublin that exciting things were happening in solid-state physics, although nobody in Dublin knew much about the field. It appealed to me, therefore, as something new and different, so I started in the solid-state theory group at the Cavendish in the autumn of 1960.

To my regret, Sir Nevill was very busy with his duties as Master of Caius College and Cavendish Professor, so he did not have the time to supervise graduate students. The small group of us in solid-state theory, including Lu Sham and Neil Ashcroft, worked with John Ziman or Volker Heine. I was with Volker and shared an office on the top floor. Those working with John Ziman were in an anteroom to his office on a lower floor near the Cavendish Professor's office, so they saw much more of Sir Nevill. Naturally we were all in awe of him as the famous professor but he also had a reputation as a notoriously absent-minded professor. There were many stories circulating about his absent-mindedness, but I am sure most of them were apocryphal.

One story I can vouch for concerns Phil Taylor, then a graduate student in the group. Phil was the only one of us who was also a member of Caius. One day Sir Nevill, on his way to see John Ziman, stopped at Phil's desk and remarked, 'Taylor, you must come to tea at the Master's Lodge.' Phil was naturally very pleased with this invitation but then days, weeks and even months went by without an actual time being set. Months later he met Sir Nevill again, who stopped and said, 'Taylor, ah yes, I was going to have you over to tea.' Pause. 'Yes, I remember now what happened. I wrote "Taylor to tea" in my diary. Later I invited Sir Geoffrey Taylor over to tea and then when he came I couldn't remember what I wanted to talk to him about!' Stories like this embellished Sir Nevill's reputation among the students as a great but rather distant figure during this period.

At one point during my student years in the Cavendish, Sir Nevill decided to find out what each of us in solid-state theory was actually working on. So he arranged appointments for us to come individually to his office to discuss our thesis topics. Now I had been assigned a problem in many-body theory by Volker Heine. This was just a few years after the famous series of papers by

John Hubbard, introducing diagrammatic techniques to the study of many-body effects in the electron gas. I was very enthusiastic about these papers and also the recently written book by Phillipe Nozières, which was my bible. So when my turn came to enter Sir Nevill's office, I launched into the diagrams at the blackboard with great gusto. Very quickly the questions started and soon my whole presentation ground to halt as I struggled to answer them. Pretty soon we came to an impasse as my formal exposition using Green's functions could not satisfy Sir Nevill and answer his questions based on his formidable physical insight and intuition. This went on for a while and I became more and more nervous and anxious as I obviously could not explain what I was doing in a clear way. Eventually he remarked that he had met John Hubbard a few months earlier and also talked to him about these new diagrams and couldn't get satisfactory explanations either. At this point I breathed a sigh of relief. After all, if the inventor of this technique could not satisfy Sir Nevill about diagrams, then surely a young graduate student could be forgiven.

It was several years later that my scientific contact with him really developed. I had spent a postdoctoral period with Walter Kohn in La Jolla, who introduced me to metal–insulator transitions, a subject which Sir Nevill had pioneered. After that I joined Bell Labs in New Jersey, and struck up a collaboration with an experimental colleague, Denis McWhan, who had developed a wholly new and large apparatus which, emitting impressive bangs and other loud noises, could simultaneously reach high pressures and low temperatures. This was eminently suited to look for the elusive Mott transition or metal–insulator transition between a Mott insulator and a metallic state as the electronic bandwidth was increased by applying pressure. It was especially elusive because, for reasons somewhat mysterious to this day, Mott insulators tend to have large energy gaps, so laboratory pressures do not have much influence. However, vanadium sesquioxide (V_2O_3) was an exception and here Denis could map out a complete phase diagram for the Mott transition, discovering an anomalous metallic state; this intrigued Bill Brinkman and me, and it led us to study how the Mott transition is approached from the metallic side. Fortunately Sir Nevill was active again in physics and a lively exchange ensued. Sir Nevill visited Bell Labs and peppered Bill and me with questions and ideas. After he left, a stream of letters followed – the first written on the aeroplane back to England. This flood of correspondence went on as Sir Nevill was wont to dash off a letter as a new idea came to him. During this correspondence the letters became more intimate and progressed from 'Dear Dr Rice' to 'Dear Rice' and finally to 'Dear Maurice'. I had to explain to my American colleagues that 'Dear Rice' was not a sign of displeasure but a decided step up the ladder from 'Dear Dr Rice'.

This topic of the vanadium oxides in the late sixties and early seventies laid the foundation for a continued close relationship between us. This was based largely on our similar approach to solid-state theory, that a theorist should not be afraid to look at complex materials because that was where qualitatively new phenomena were to be found. This should be the frontier rather than the refinement of the theory of simple metals and semiconductors. Indeed, I watched with great admiration how he could make sense of complex materials, such as amorphous semiconductors and ionic liquids, where I feared to tread. His books were always worth consulting and stimulated one to think about a series of unusual and fascinating materials.

His last visit to Zürich was some years ago. As the Cold War was coming to an end, the possibilities for many Russian physicists to travel to the West finally opened up. So in the late eighties I could arrange for Alex Efros and Boris Shklovskii from Leningrad to spend a month in Zürich as their first visit to the West. This in turn attracted Sir Nevill to Zürich, since their highly original work on the Coulomb gap in a random localized insulator had led to important modifications to Mott's famous $T^{-1/4}$ law for variable-range hopping in such systems. It was a very enjoyable and stimulating time for all concerned. He reminisced about his youth – the period after the discovery of quantum mechanics was, he said, a great time to be a young physicist because the field was wide open and there was so much to do, and the older physicists were out of the picture because of their great difficulty in understanding the new quantum mechanics. He talked sadly about Klaus Fuchs, the notorious atom spy, who had been his student. He was, he said, reluctant to go to Eastern Europe because he did not want to meet Fuchs; 'I wouldn't know what to say to him,' is how he put it. We had discussions covering a wide range of materials which were the object of Sir Nevill's enormous curiosity. For me it typified his insatiable desire to understand materials in all their aspects, which inspired him to the end. I will always remember his gracious manner and unfailing courtesy to everyone he met.

PROFESSOR T. M. RICE joined the Cavendish Laboratory as a graduate student with Volker Heine in 1960. After graduating in 1964 he went to the United States and spent fifteen years on the staff of Bell Laboratories. In 1981 he took up his present position as a professor in the Institute for Theoretical Physics at the Eidgenössische Technische Hochschule in Zürich, Switzerland.

Can You Hear Me in the Back?

Nic Rivier

*Von Sommerfeld hab' ich Optimismus gelernt, von den Göttingern die
Mathematik, von Bohr die Physik*

Heisenberg

I learned from Professor Mott the enjoyment and the fun of physics. Here are
a few examples.

I arrived in Cambridge in late November 1964, armed with a degree in
physics from Lausanne but with only ten lessons in English by the Assimil
method. For the reader not familiar with this breakthrough in pedagogy,
Assimil teaches English by making you read aloud useful sentences like 'my
tailor is rich' with the help of an encoded pronounciation straight out of Dr
Caius (from Windsor, not from Cambridge). For example: 'If dere be one or
two, I shall make-a de turd'. By lesson 10, we had encountered the future
tense, but only just, so I was fully prepared to discuss my future with Sir
Nevill.

My letter had probably indicated the level of my competence, for it had
caused a beginning of panic in the theory group at the Cavendish. Ian Aitchi-
son had been 'volunteered' to greet me and to take me to lunch before my
appointment with Professor Mott, on the grounds that his knowledge of
French was the most recent as he had just returned from a fortnight in Orsay
(learning French by osmotic pressure). The lunch, incidentally, was very pleasant and instructive; I learned that the pronunciation of Thouless caused
almost as many problems to the English as to the French.

So I find myself, not too confident, in Professor Mott's office. Prof. tells
me, with extreme clarity and the diction of a trained actor, 'Monsieur Rivier,
would you mind very much if we spoke in French. I would like to take this
opportunity of practising my French.' An offer I could not refuse.

At that time, the Solid State Meeting of the Institute of Physics was held
every year in Manchester just after New Year. Professor Mott had been
invited to talk on one of his recurrent interests, why mercury is a liquid, but
between his accepting the invitation to speak and the lecture, he had realised
that he did not have much new to say on the subject. Yet the lecture was
memorable. He spent the entire half-hour in writing down the Schroedinger
equation, with a timing and wit which kept the audience enthralled. Scenery: a
blackboard. Technical props: one microphone on a stand frontstage, the other

around Mott's neck, connected with a very long cable. Audience: packed but sleepy at first (it was just after lunch).

Mott (*two metres away from the fixed microphone, very quietly*): Can you hear me in the back?
Rustling from part of the audience which cannot hear; other part snoozing.
Mott (*now five centimetres from the fixed microphone, very loudly*):
Can you hear me in the back?
The audience wakes up, in phase.
Mott: To understand metals, one must write the Schroedinger equation.

He then turns around by π radians clockwise, goes to the blackboard, to write one half of an aitch bar. He then faces the audience (by making a second turn of π clockwise) to give some interesting aside on the Schroedinger equation, metals and/or mercury. Another rotation of π clockwise, and we have the full aitch bar and a bit of the em: 'Now, the mass is interesting ....' And so on, always clockwise, until he arrives at the vee, after a total of seven and a half clockwise rotations, with the cable of his microphone wound around him seven and a half times in a nice, regular helix, and the audience in suspense.

A comment on the potential is given after a counterclockwise rotation of π. He goes on to write the equation to the end, with exactly six interesting asides between counterclockwise rotations, to finish exactly unwound on the half-hour, with his last comment on the last wiggle of the final psi. Deserved standing ovation.

As thesis supervisor, Professor Mott had to arrange the viva for my PhD. It turned out that the three persons involved, the two examiners and myself, could not all be on the same side of the Atlantic at the same time. On the phone with the Secretary of the Board of Graduate Studies, Prof. explored the various possibilities, such as having two separate vivas (one with the internal examiner and a witness, the other with the external examiner and a witness, and presumably a conference between the examiners). He was told that this was against Regulations, but that the examiners could decide, after reading the thesis, that a viva would not be necessary. Prof. instantly arrived at the solution, which is the product of a great diplomatic mind: I was going to have two separate vivas, after which the examiners were to recommend jointly that a viva was not needed. In the same vein, one of the Caius legends was that Mott, as Master, had suggested that the dog of a Fellow should be officially called a cat (dogs were not allowed in Caius).

Perhaps the nicest touch, and certainly one of the most welcome, by the Motts, was the ladder in Caius. To get into College after the gates were shut, one had to climb a wall bordering the Master's Garden. There was always a ladder left on the other side to help us down. Of course, Lady Mott claimed that it was only to prevent her chrysanthemums getting damaged. The gesture was just nice; the explanation beautiful. Both were typical.

Thank you, Professor and Lady Mott, for the lessons and for the memories.

PROFESSOR NICHOLAS RIVIER was born on 5 August 1941 in Lausanne, where he was educated. When he was a research student in Cambridge (1965–1968), Nevill was his supervisor. In 1970 he was appointed a Lecturer at Imperial College, London, and subsequently became a Reader in Theoretical Solid-State Physics. From 1993 he has been a Professor at the Université Louis Pasteur in Strasbourg.

Excursions into Politics

Tam Dalyell

'You really must find me more academic heavy-weights, apart from Pat Blackett!' Thus, Dick Crossman, within a week of his appointment by the then newly elected Leader of the Opposition, Harold Wilson, in February 1963. To explain, I was eight months old as a Member of Parliament and had just been made Secretary of the Labour Party Standing Conference on the Sciences, with responsibility for working to Crossman and helping to put together 'Two-Way Traffic in Ideas' Conferences, up and down Britain, with the purpose of spatch-cocking together a science policy, in circumstances where the Labour Party policy cupboard was pretty bare, at the time of Hugh Gaitskell's untimely death in January 1963.

As an undergraduate in Cambridge when Nevill arrived from Bristol, I had attended his contribution to a series of open lectures for those of us who were not physicists. Like many of my generation, I had been quite simply captivated by the capacity of this formidable presence to convey important ideas. Others who participated in this series were Sir Edward Bullard, a Fellow of both Clare and Caius, Director of the National Physical Laboratory at Teddington, no mean performer, who after forty years have elapsed, I still recollect enthralled us students with his geodesic domes; and Otto Frisch, FRS, Jacksonian Professor, who introduced us to particle physics, and waiting until his huge audience in Mill Lane was expectantly hushed, memorably began, 'When my Auntie Lisa was splitting the atom ....' Frisch's auntie was, of course, Lisa Meitner. Nevill did not indulge in such dramatics.

Nevill was to the point. No embellishments. It came through to us students that what he really cared about was that the non-physicist undergraduate should grasp an inkling of what modern solid-state physics was all about. And this indeed was one of his qualities. He really cared about conveying the message. He, as messenger, and his reputation, were of no account. It was the message that mattered. He was a man of no self-importance. The issue was the important thing. Such an attitude is somewhat rare among those involved in public life, or even dare I say, in academia.

This was the background against which I said to Crossman, 'Try Nevill Mott. I doubt if he will have much to do with us, but he just might!' Crossman did. I was not present at their dinner *à deux*, the date of which I had arranged. But I heard Crossman's version an hour later, as I was his lodger in the upstairs rooms of his London home at number nine, Vincent Square.

This great brown bear of a man was very suspicious of me at first. 'Could a New College, clever Wykhamist don in Greats and philosophy really be interested in science?' he was obviously wondering. But, fortunately, Madam Prunier [Prunier's was Crossman's favourite eating place] was excelling herself and as the dinner went on, Mott thawed. He is very heavyweight indeed. He is brimful of ideas. He is obviously concerned. I don't think, for a moment, he votes Labour, though he told me that he did so in 1945. But he'll help. He's potentially a great catch!

Years later, Nevill recalled the self-same dinner to me:

I rather warmed to your Mr Crossman. It is rubbish for you to suppose that I was suspicious because he was a classics don – if a chap with a first-class honours in Greats wanted to come to my lab and was keen to train in physics, I'd take him like a shot. No, I was wary because I thought that he might be too good to be true. But, I thought to myself, if a Shadow Minister and prominent member of the National Executive Committee of the Labour Party is prepared to give me time (and an excellent dinner) I am disposed to help.

And help he did. He made a fourfold contribution to the Bonnington Conference, so called after the Bonnington Hotel in Southampton Row, in whose London basement the final conference of thirty-four Two-Way Traffic in Ideas took place in the presence of the Leader of the Opposition and several members of the Shadow Cabinet.

First, and perhaps unexpectedly, – Nevill very often said what was unexpected from a Cavendish Professor – he said that any Labour government must give priority to schoolteachers in general and maths and science teachers in particular. (Indeed, he argued the case for special rates of pay for maths and science teachers, who at that time could be tempted into industry by salaries two or three times above those which they could command in schools.) Though doubtless unintentional, this emphasis on school education placed him in the pews of the realistic rather than in the pews of the airy-fairy, in the minds of Labour politicians.

Secondly, he championed the Robbins proposition that teaching and research should go hand in hand. Along with Vivian Bowden, then Principal of UMIST, later to be Minister of Higher Education, October 1964 to March 1965, until he sacked his (Permanent) Secretary, Sir Bruce Fraser, without so much as telling Michael Stewart as Secretary of State, let alone the Prime Minister, Nevill expressed scepticism about the value for money of the resources at the military research centres of excellence, such as RSRE Malvern.

Nevill and Vivian thought people who did not have teaching obligations 'withered' into a comfortable life – something Nevill was never himself to do, even in extreme old age, when he would propel that huge frame from his home in Milton Keynes down to Westminster and up the stairs to the dining-room and committee rooms of the House of Commons. He was never to grow old other than in the physical sense. He ordered me – yes, Nevill's hints in 1963–64 were my orders, to go to the Services Electronics Research Laboratory (SERL) at Baldock, near Cambridge, to see their work on gallium arsenide. A week later, Nevill asked, 'Did you not sense that with those facilities which you saw at SERL the people you met would benefit if they were part of a university with teaching obligations?' I did. (I believe to this day that Nevill and Vivian were right and that far more could have been done to turn the RSRE into something equivalent to the Technical High School in Zürich or other great Continental postgraduate institutions.)

His third contribution was a passionate concern about the value of fundamental science, where the inquisitive could get resources to allow them to venture into the unknown, even if there were no commercial advantage on the horizon: 'You cannot tailor science to profitable ventures!' Resources must be made available for good people and for uncomfortable people. Among those whom he had in mind were Sir Ernst Chain, FRS, with his ground-breaking ideas on chemical microbiology, and his professorial colleague of Bristol days, the Communist C. F. (Cecil) Powell, FRS, pioneer of nuclear physics in photographs.

But Nevill's fourth contribution was the abiding and ever present concern about the limitations on resources. He endeared himself to Jack Diamond, then Labour MP for Gloucester, later to be Chief Secretary to the Treasury, by making it clear that he recognised how science and education were not the only departments to make priority calls on the resources available to a Labour government. Investment in public transport was high up Nevill's priorities. And he was among the first to warn that the demand on National Health Service finances would rocket with the kind of operations that his friend, now Sir Roy Calne, was undertaking with kidneys and livers up the road at Addenbroke's Hospital.

This led him to preach to us about what he called 'the mechanism of shrinkage'. If there was to be expansion, particularly in areas of interdisciplinary development, such as molecular biophysics, then there also had to be contraction. Rightly, or perhaps wrongly in retrospect, he cast a beady eye on the expensive schools of anatomy, perceiving them to be outdated. He was also no friend to the extremely expensive demands of some of his colleagues in nuclear physics, who were insensitive, as he saw it, to the relatively small sums required by the solid-state physicists. More particularly, he asked, 'Who was to incur the fury of worthy academics who had devoted their worthy lives to their worthy research and found that it was cut off at the end of the day?'

This was a question, like a number of Nevill's questions, which has never quite been answered, since any answer is too awkward to formulate, other than in the vaguest terms of mutterings about peer-group decisions. Nevill would say that he understood perfectly the human misery for decent scientists that the mechanism of shrinkage could cause, in the killing-off of research work and the process of evaporating university departments.

To be candid, I'm not sure that he did. He was a person of such intense curiosity and inquisitiveness about the world himself, not only in science matters, but ever increasingly in religious matters, that I do not think he appreciated the sadnesses inflicted on other people of narrower minds by the pruning of what he would see as dead wood, and they would see as their research babies.

In October 1964, like most of the other academics participating in the Two-Way Traffic in Ideas, Nevill was nonplussed, and indeed angry, that Crossman was not given the science and education portfolio, which went instead to the ex-classics teacher Michael Stewart, later Foreign Secretary. It looked like, and indeed was, the squandering of a lot of busy people's time, which had been given to the Labour Party unstintingly in opposition on the understanding that they were contributing to a future science policy and that Crossman would be in charge, even if he did not accept all their ideas.

When I saw him, in November 1964, I was Crossman's Parliamentary Private Secretary, in the Ministry of Housing, at the start of the sagas with the formidable Permanent Secretary Evelyn Sharp, the well-remembered Dame of the Crossman Diaries.

'Why did Crossman not get the job as Secretary of State for Education and Science?' Candour was the only way with Nevill. First, I told him I would give him the ostensible reason put around by Harold Wilson as Prime Minister. 'He says he needed Crossman's volcanic energy to achieve something tangible in housing in a matter of months before he has (with a majority of three) to go to the country again.'

'No, no, no,' said Nevill. 'Now, let me have the real reason.' And so I explained.

Dick, tired at the end of a bad day, can be about the rudest, most boorish man in England. He was quite dreadful to the National Union of Teachers representatives of the Nursery Schools when they came to see him in the late summer. They were enormously offended. Snubbed, and with a terrible sense of grievance by being made to look foolish, they complained bitterly to the National Union of Teachers MPs, particularly to ex-headmaster Ted Short, later to be Wilson's trusted Chief Whip, and ex-schoolmaster George Thomas, a particular friend of Wilson, later to be Speaker, who went to the Leader of the Labour Party and said, 'You cannot put that awful bully Crossman into Education!' Wilson demurred, but felt it prudent to give in.

After Crossman died, I checked with Harold Wilson exactly what had happened, and he confirmed that this had been the case. Nevill, on hearing this explanation, simply sighed quietly, 'A pity, but I understand.'

Unlike many of the participants in the Two-Way Traffic in Ideas, Nevill did not sulk and wash his hands of the Labour government. He was never a fair-weather friend; he turned out to be a foul-weather stalwart in terms of constructive advice. (Though I suspect that his personal vote did not go to the Labour Party in 1970.)

To the best of my knowledge, he had no rapport with Michael Stewart (I am fairly confident of this as I was asked to contribute Michael Stewart's entry to the *Dictionary of National Biography*), after Vivian Bowden had been discarded in March 1965. But he did contribute to the thinking of Anthony Crosland, Stewart's successor, and his senior civil servant in charge of the work on the Binary System, Toby Weaver. (Nevill, I noticed, took immense pains about giving the civil servants their place and treated them as human beings. In turn, they took him exceedingly seriously, and he was all the more effective on that account.)

Nevill's basic belief, shared by Crosland as Secretary of State, was that any boy or girl who had even a slim chance of success at University, should be accorded the opportunity, both from the nation's point of view, but more importantly for his/her self-esteem. The rights of the individual were important to Nevill.

Accustomed to civilised Tory Education Ministers such as Sir David Eccles and Sir Edward Boyle, whom he admired, he had hopes of Mrs Thatcher, with her science degree, as Education Secretary. These were dashed, and he would ask, both during her term as Education Secretary (1970–1974) in the government of Ted Heath and as Prime Minister (1979–1990), 'How can a pupil of Dorothy Hodgkin, FRS, do such things?'

Finally, Nevill Mott was quite simply a good man and an extremely nice one. Nobody could have been a more charming host to my wife, Kathleen, in talking to her properly and without a trace of condescension. Edmund Dell, later to be a Labour Cabinet Minister, recalled that Nevill had given us an interesting and extremely worthwhile weekend in Caius, which he remembered thirty years later. His kindness sticks in the mind, along with his interest in other people.

The last time I saw Nevill was when he came to lunch with me in the House of Commons during one of his many busily filled days in London, when well into his eighties he had trundled down from Milton Keynes.

'Nevill, what would you have been had you not been an academic?' I asked. 'Well,' he said, 'I would have been a calamitous politician.' Alas, that might have been true, as he looked for the good in people who disagreed with him.

'How about industry?' I pursued. 'Yes, certainly!' He had a good deal to do with the setting up of the now thriving Cambridge Science Park.

'Or the Church? A bishop?' I continued. 'I think not,' he said.

'Or Scotland Yard, what do you say to that?' He evinced interest.

When Deputy Premier Rudnev asked Harold Wilson to send a Labour Party delegation of scientists and politicians to the USSR in March 1964,

Nevill was the first choice. He couldn't get away. Instead, he suggested David Schoenberg, FRS, a Russian-speaking friend, Fellow of Caius and Director of the Mond Low Temperature Laboratory. After we returned, he put me through an interrogation about the Russian structure of science such as I have seldom been through.

Others in this volume are better placed to assess his professional work, but I shall remember Nevill for his demonic mental energy in formulating the important questions and persisting, ever returning to the crux of the matter, until he got an optimum response. Curiosity driven, he was a superstar detective.

TAM DALYELL, MP, has represented the West Lothian/Linlithgow constituency since 1962; he has written a weekly column for New Scientist *since 1967 and was given an Honorary Doctorate of Science from the University of Edinburgh. He and his wife, Kathleen, were personal friends of Nevill Mott from 1963 until his death.*

Tripos Reform and Invention

Richard Eden

Until 1850, candidates for honours in classics in Cambridge were required first to have obtained honours in mathematics. Although this requirement had been long forgotten when Nevill Mott came to Cambridge in the 1920s, the prestige of mathematical studies remained and the Mathematical Tripos was the main route to research in theoretical physics, a route which was taken by Nevill himself and all his predecessors as Cavendish Professor, except Rutherford. Colleges provided few facilities for research students, and those in theoretical physics generally worked in isolation from each other and from the Physics Department itself. Mott was a student in St John's College, where Dirac was then developing the foundations of quantum mechanics. But there was no useful interaction between them, and soon after starting research Nevill left Cambridge for Copenhagen, where Niels Bohr had created a group which was a prototype for modern interactive research in theoretical physics. Except for an increased number of research students, this early environment for theoretical physicists in Cambridge was not much changed when I was a research student after the war. Our common experience made me a natural ally of Nevill's in his aim to facilitate the development of theoretical physics research groups in Cambridge on the Copenhagen model, or – in my case – the American model.

In an early move to strengthen theoretical physics in Cambridge, Nevill invited Hans Bethe to come as visiting professor during 1955–56. During that year I was in Manchester but was collaborating with Bethe in research on the nuclear many-body problem, not dissimilar then to emerging work on solid-state physics, and this may have contributed to my receiving an invitation to return to Cambridge to the mathematics lectureship that became available when Abdus Salam left for Imperial College at the end of 1956. In the summer of 1957, following the decanting of molecular biology from the Austin Wing of the Cavendish into a hut in the courtyard, Nevill provided me with two offices in part of their former laboratory, which allowed me to find space for a research assistant and two research students. A year later, Nevill also helped with arrangements for Hamilton and Polkinghorne to have offices with their students in a temporary hut on a car park where the Holiday Inn now stands. We were subsequently brought together in the summer of 1960 on the new top floor of the Austin Wing, which was built with unusual speed in consequence of Nevill's decision to use the residue of the Austin endowment for this purpose, though in the event, additional finance was required from central funds. The whole of this new facility was for theoretical physics, about two-thirds being used by staff and students (numbering about thirty) from the Department of Applied Mathematics and Theoretical Physics (DAMPT). Most of us were then working on quantum field theory, particularly the S-matrix and dispersion theory, but there was also some work on collisions of elementary particles of direct relevance to experimental work in the Cavendish. The other third of the new facility was occupied by Ziman and Heine and their research students in solid-state theoretical physics, all being members of the Physics Department.

With the objective of relaxing the artificial separation in theoretical physics between the Mathematics Faculty and the Physics Department, in the autumn of 1957, following discussions with Nevill, I had sought to create a two-year tripos in theoretical physics. DAMPT was not formed until 1959, and as a member of the Faculty Board of Mathematics I had no difficulty in getting some support for this proposal. Although he was an ex officio member of this board, I have no recollection of Nevill ever attending its meetings. However, when the proposal reached the stage of discussion in the university at large, there was impassioned opposition from Sir Charles Darwin, then Master of Corpus, who thought it would be too narrow an education. I did not agree and neither did Nevill, but he advised that the arguments would be less clear to the members of the university at large if it came to a vote, and might even be sunk by the General Board of which he was a member. We therefore switched tactics and, without difficulty, established the Mathematics with Physics Tripos, Parts I and II, which allowed mathematics students to be examined on lectures in physics given for the Natural Sciences Tripos, instead of applied mathematics lectures given by mathematicians.

In the Lent term of 1957, during lunch in Newnham following a meeting of mathematics examiners, I had described how a prospective student from an underprivileged background had done well in the entrance scholarship examination but had declined an offer of a place in Clare because he had never studied Latin and would be unable to qualify in the time available. Ray Lyttleton suggested that we should initiate a campaign to abolish the Latin requirement for Cambridge admissions. Our initial step was to consult Mott for support by publicity within the Cavendish, later extended to other science departments. Nevill was not much involved in the campaign, but his support and advice in the early stages was an important factor in the decision to go ahead. In 1958, following a massive vote in the Regent House, it was decided to drop the Latin requirement. The objective to help the underprivileged was partly sabotaged in the detail by a requirement of two languages as an alternative, but the second language was dropped five years later, on the initiative of the Colleges Tutorial Committee.

The idea of a theoretical physics tripos that had been sunk by Darwin when we approached it from the side of the Mathematical Tripos was salvaged in the context of the Physics Tripos during the early 1960s by Mott, Heine and myself. Our objective was to provide a route to research in theoretical physics within the Cavendish. We did not plan to compete with Part III of the Mathematics Tripos, and our aim was to encourage students to use quantum theory in applications to physics rather than to learn about more formal developments of the theory. The new Theoretical Physics Tripos was in place by the summer of 1964 and provided a good reason for my own transfer from DAMPT to the Physics Department to teach some of the new courses. The first examination was in 1966. Since the teaching and examining were entirely within the Physics Department, it was relatively easy a few years later to introduce a middle option, a mixture of theory and experiment in Part II, which was retitled Physics and Theoretical Physics. The new theoretical physics courses proved popular, particularly with the better students, and many of them subsequently proceeded to research in one of the experimental groups in the Cavendish. Among the latter was Richard Friend who, as Cavendish Professor, is one of the successors to Nevill Mott.

RICHARD EDEN, OBE, is Emeritus Professor of Energy Studies at the Cavendish Laboratory. Prior to taking up interdisciplinary energy studies in 1972, he was a Reader in Theoretical Physics researching in quantum field theory, nuclear theory and S-matrix theory. He is an Emeritus Fellow and an Honorary Fellow of Clare Hall.

Metal Physics at the Cavendish: a personal appreciation

Peter Hirsch

Nevill Mott became Cavendish Professor of Experimental Physics in 1954. At that time I held an ICI Fellowship at the Cavendish Laboratory, in W. H. Taylor's Sub-Department of Crystallography. I spent part of my time extending the work on the structure of coals and carbons by X-ray diffraction, on which I had been engaged since 1950, but the fellowship enabled me also to get back to my PhD field, work hardening in metals. In 1954 I initiated with my student, M. J. Whelan, a programme to explore the possibility of observing dislocations directly by transmission electron microscopy of thin metal foils. Nevill Mott's arrival at the Cavendish at this time was a most fortunate coincidence. He gave us tremendous encouragement, and arranged a grant for me from the Ministry of Supply from 1955, when the ICI Fellowship came to an end. Nevill's interest in and support of metal physics in the Cavendish Laboratory were crucial to the growth of this research area in subsequent years. Following our success with the electron microscope work, Nevill provided us with space in the old Cavendish for our microscopes and other facilities in the Austin Wing for our rapidly expanding activities. We split off from the Crystallography Department and revived the Metal Physics Group, which had become defunct when Orowan left for the United States. Nevill also supported us with university posts; by the time I left in 1996, Nevill had obtained a readership for myself, M. J. Whelan was an assistant director of research, A. Howie was a lecturer, and L. M. Brown and A. J. Metherell were demonstrators.

Just as important, if not more so, was the intellectual stimulus provided by Nevill. It was marvellous to have a head of department who had prime research interests and was an authority in one's field. We had countless discussions on work hardening, creep and fatigue. Nevill would always raise critical issues and ask the difficult questions, and these meetings used to act as a stimulus for further experiments. The interpretation of the experimental results was also much influenced by his theoretical models.

Our work on dislocation distributions and strength of cold-worked metals was much influenced by Nevill's suggestion that the proportionality of the temperature-dependent and temperature-independent parts of the flow stress, as found experimentally by Cottrell and Stokes, is due to the elastic and short-range interactions originating from the same 'forest' dislocation. This is the basis of the forest theory of flow stress. Similarly, our researches on jogs, flow stress at high temperatures, and fatigue were stimulated by some of his ideas.

Sometimes during the scientific discussions in his office I found myself distracted by his efforts to light a cigarette. I got the impression that he wanted to reduce smoking. To achieve this, he limited himself to packets of ten cigarettes, and there were never any matches. So a one-bar electric fire was switched on and Nevill would use it to light a large paper spill; this would promptly burst into flames then required considerable effort to extinguish it. The whole procedure appeared somewhat perilous.

Nevill attracted all the good and the great in the field, and there were many visitors who took part in the discussions, among them Friedel, Crussard, Seeger, Nabarro, Eshelby, some of whom spent extended periods in the Cavendish. Nevill was often very enthusiastic about one or other aspect of mechanical properties, particularly in the 1950s, and he expected others to be similarly committed. In the summer of 1956 he had organised a meeting at fairly short notice at which Tony Kelly was supposed to give a paper. Unfortunately, the meeting coincided with Tony's wedding; Nevill nevertheless expected Tony to fit the meeting in somehow!

There are many stories about Nevill's rather endearing absent-mindedness and idiosyncrasies. I remember attending a lecture he was giving in the old Maxwell Lecture Theatre. In those days there used to be a large epidiascope on a bench in the front, covered by a black cloth to keep out the dust. At the end of the lecture he picked up the cloth, mistaking it for his gown. His secretary, Miss Glasscock, returned it in due course – she had done it before!

Long after he had 'retired' and we had moved to Oxford, he visited us on Boxing Day, carrying with him the manuscript for the second edition of Mott and Davis, to be delivered to the Oxford University Press. He was much put out to find the OUP was closed on Boxing Day!

Nevill provided me with scientific inspiration and support when it was needed, and I greatly appreciated his sound advice. My wife and I much valued the warm friendship of Nevill and Ruth. I was touched when he came over to Oxford in 1992 to the dinner to mark my retirement, and made a charming speech. I owe him an immense debt of gratitude. He gave me and the other members of the Metal Physics Group a unique opportunity to develop our scientific field. It was indeed a great privilege to have worked under his stimulating leadership at the Cavendish for so many years.

SIR PETER HIRSCH, FRS, was born in Berlin in 1925 and educated at Cambridge University. He was Assistant Director of Research, Lecturer and Reader in Physics at the Cavendish Laboratory 1957–1966. He was Professor of Metallurgy at Oxford University 1966–1992 and part-time Chairman of UKAEA 1982–1984. He has researched on plastic deformation and electron microscopy of defects in crystals and in 1956, with Whelan and Horne, he made the first observations of dislocation motion in thin metal foils by transmission electron microscopy.

Selected Encounters

Mick Brown

In 1960 I arrived in Cambridge to work with Peter Hirsch as a postdoc on an Admiralty grant concerned with the deformation of metals at high temperatures. In the Department of Physical Metallurgy at Birmingham, where I had recently completed my PhD, one could see that electron microscopy would revolutionise the study of solids, and transform speculations about defects into observations which could account for behaviour – a kind of new engineering, building with mesoscale structures. At that time, you had to be in Cambridge to do it. Nevill Mott's name was known to me through what we called the 'Mott and' books: Mott and Gurney, *Electronic Processes in Ionic Crystals*; Mott and Jones, *The Theory of Metals and Alloys*; Mott and Sneddon, *Wave Mechanics and Its Applications*. I had read the first two, with limited success at gaining understanding from them, and had tried without any success to read the third. I had not even seen Mott and Massey. To my surprise, I found that the theory of electron scattering from atoms, so crucial to electron microscopy, was based on Mott's formulae, derived in the latter book.

It was therefore with excited anticipation that I found myself in the seminar room of the Austin Wing in the old Cavendish, awaiting a seminar from Nevill Mott on the electronic structure of liquids. The first thing I noticed was that the room was full – you had to scramble for a seat. The second thing was the electric blinds – at a word from the lecturer, the blinds would slowly trundle down, blacking out the room for the projection of transparencies. And the third thing was the lecturer himself – very tall, shuffling through a pile of papers on the lecture bench, and writing awkwardly on the blackboard a list of Poisson's ratios of metals; there were certain crucial gaps in the table. There was an introduction by Brian Pippard, who mentioned the balance between research and teaching in universities. Nevill started to speak, explaining that there was a difference of opinion concerning teaching: Nevill thought the mixture of research and teaching should be about 50:50, whereas Brian thought it might be nearer 30:70. Having reduced their difference to a mere matter of emphasis, Nevill began his seminar. His voice ranged in an alarming manner from a great boom to an inaudible whisper. He seemed to be drawing ellipsoidal Fermi surfaces on the blackboard. He called for the blinds to be opened – he had no transparencies – and the loud electric motor drowned him out; the blinds opened with agonising slowness. He shuffled through his papers, seeming to have lost a crucial sheet. The Fermi surface on the blackboard now had fuzzy edges. The crucial sheet of paper still eluded him, even after another shuffle of papers. The pile of papers was now distributed chaotically over the lecture bench. The packed audience became restive. At that point, I got the giggles and I left the lecture in confusion and shame. Afterwards I asked Frank Nabarro, a visitor to Hirsch's group at the time, what he thought of it. 'Brilliant!' he replied. 'I wish I could give lectures like that.'

The next encounter occurred when I was a candidate for a W. M. Tapp Research Fellowship at Gonville and Caius College. I was about to be interviewed, Nevill as Master chairing the interview panel. He took me out to lunch the day before the interview and asked me what I was doing and what I intended to do. I referred to his work with Nabarro and how electron microscopy made it possible to see very directly the local strains around precipitates in solids, as well as dislocation arrays. I hoped, I said, to understand the mechanisms of fatigue and work hardening in metals. I remember very little of what he said during the lunch. But as we got up from the table, he looked me clearly in the eye and said, 'I think it better if at the interview you don't mention your connection with my early work.' During the interview itself, he asked no technical question, but left it to others. I was asked whether or not my work would be better carried out in a metallurgy department or a physics department. I replied that I did not know, but perhaps it could be done wherever people were interested in these questions; certainly the Master had contributed famously to them. The matter was dropped at that point. As a result of the interview, I became a Research Fellow and launched on an academic career, although it certainly didn't feel like a launch at the time, more like another temporary job. There is little doubt that Nevill's political astuteness was crucial to my success.

Gonville and Caius hold opulent dinners in the Edwardian style, and at my first one, the Perse Feast, Nevill rose from the high table and wandered among the other tables, accompanied by lesser college figures. He seemed very tall; he wore only an MA gown, which made him seem taller, and his leonine head was set against the background of the stained glass and panelling of the dining-hall. He asked younger fellows in turn about their families and their activities; one of them, sitting near me, was about to run as a would-be Tory MP. It was a very grand scene, and Nevill positively relished being the grand figure at the centre of it. His portrait, which until very recently hung in the hall, is a good likeness. An abstract expressionist display, like a halo in the background, shows the fireworks which seemed to move with him as he progressed around us in the hall. The expressionist halo also conveys perhaps the scientific fireworks, which are, however, much more subtle and require effort to appreciate.

I had (and still have) a colleague and friend who is my double, John Leake, then in the Crystallography Group, now in the Department of Materials Science at Cambridge. One day he told me a story of meeting Nevill, who asked him, as he often did, 'Tell me what you're doing now.' In the course of John's reply, he noticed the well-known glazed look overtake Nevill's eyes. When John had finished, Nevill said, 'That's odd, Brown; you weren't doing anything like that last time I talked to you!' For many years afterwards, I believe that Nevill found it difficult to know which of us he was addressing; his memory for names and faces was notoriously unreliable. Yet he could with sharp accuracy recall people when there was a real need to do so.

A group of us in that period were keen enthusiasts for the abolition of nuclear weapons and participated in the Aldermaston marches organised by the Campaign for Nuclear Disarmament. Nevill, as a Pugwash member and proponent of the 'hot line', the instant telephone communication between the President of the United States and his opposite number in the Kremlin, spoke often on the subject. His views, while liberal by the standards of the day, did not extend to unilateral nuclear disarmament as advocated by Bertrand Russell and other heroes of the CND. He very much encouraged young members of his department to get involved, and I remember that a paper we wrote proposing the establishment of a 'peace research institute' was duplicated and distributed using equipment in the secretary's office in the Cavendish. Nevill read our paper, and had only one comment that I remember: we had misspelled his name, Neville instead of Nevill. It is a very common error. Shortly after that, he encouraged me to attend a Pugwash conference in Poland; I had no paper to prepare, just to take part in the discussions. Much to my surprise, it was hard work, constant social contact with a variety of nationalities and scientific disciplines, and the strain of diplomacy. Nevill saw these activities as part of the wider educational objectives of his Physics Department, a way of leavening the loaf. After CND fell out of fashion, he encouraged lectures on science in society. One wonders if under the pressures of today it is possible to take such a liberal view; the Research Assessment Exercise and the Teaching Quality Assessment make it hard to support non-examinable, non-publishable activities.

With strong encouragement from Peter Hirsch, I applied for a demonstratorship in the Physics Department; Nevill said to me, 'Ah Brown, you want to become one of us? Very flattering. What are you doing now?' I described work on dislocation line tension and the shape of threefold nodes observed in the electron microscope. 'Very interesting,' he said, 'I liked your earlier work better.' I often felt that Nevill wanted me to be doing something either better or more interesting than what I was doing at the time.

After cutting my teeth on a first-year course on the properties of matter, he called me into his office and asked me to take over his lectures on quantum mechanics, at that time a hot bone of contention between the solid-state physicists, who taught it as wave mechanics, and the particle physicists, who taught it as matrix mechanics. Of course, his little book on wave mechanics was widely used, but his lectures were not that well received by the students, as I knew from supervising their studies. I was very flattered, and agreed to give the course at once; he muttered under his breath as I left the room, 'Good, that will bring fresh life into it.' My first year of presentation felt like a disaster; due to pressure of time I omitted spin altogether and one of the senior particle men scolded me over tea, 'You're sending people into the world totally uneducated, ignorant of spin!' Nevill asked me what I had told them about the periodic table, and I said what I had done. He looked at me quizzically, and said, 'Well, the students don't know you said that.'

At about that time, shortly after we moved to the new building, I was in his office for some reason to do with research support. Among the litter of papers on his desk, I noticed an absolutely blank sheet, right on top of the pile, blank except for a single phrase: 'The hopping electron' and under it an equation. The sheet was so visible that I had the strong impression it was a deliberate plant on his part, to express graphically his sense of being caged in by bureaucracy, countered by the adventure offered by a fresh blank page. I asked him if, when he retired, he could again teach some of our first-year students at Gonville and Caius – they badly needed inspiration. He explained how flattered he felt, but refused, 'I don't want to be obliged to *anyone* after I retire.'

At another stage, shortly before he retired as Cavendish Professor, he revealed that he had read our papers on work hardening in composites and dispersion-hardened metals. He approached me in the canteen and asked, 'Tell me very simply, one sentence only, how you have stabilised the internal stresses against unloading.' I gave him the answer that it was because the dislocations were in loops at obstacles and couldn't run back. 'Ah!' he said, 'Yes, of course.' I felt on top of the world. He read my papers! In fact, he spent a great deal of time reading and absorbing other people's papers.

Two other memories regarding papers come to mind. In his little book on wave mechanics there is a problem concerning the existence or not of bound electronic states in different three-dimensional potentials. It seemed to me at the time that these questions really concerned localisation and the metal–insulator transition, and I thought through the problem as it might apply to cylindrical potentials appropriate to a dislocation. I wanted to discuss it with him, but the discussion did not take off; he dismissed me with the claim that it couldn't be that simple. Only later did I read more about metal–insulator transitions and come to understand more about the complexity of the topic. On another occasion, with first-year students in the practical class, we did simple experiments on the flow of electricity through mixtures of insulating and conducting spheres to study the percolation threshold in a random array. I thought the work was interesting enough to publish in *Philosophical Magazine*, and submitted it; however, the paper was rejected by Nevill himself on the grounds that he didn't want the journal to be filled with such simple stuff, meaning I think that there would be many follow-up papers whose publishability would be difficult to decide if mine were published. Eventually it came out in *Physics Education*, and, so far as I know, it has had no successor. In fact, better undergraduate experiments were subsequently carried out by other teaching staff; the measurements I reported were flawed because the mechanical mixture was not really random. But I include these stories to show that Nevill was very concerned to preserve or perhaps to demonstrate respectability.

Nevill certainly felt that band theory, as we now teach it in undergraduate solid-state physics, is overemphasised. In fact, his ideas are easier to grasp and to teach than details of band theory and the power laws which arise in the theory of electron transport at low temperatures. Yet his ideas relate to phenomena readily displayed by materials in our familiar environment. In my recent attempts to teach the subject, I failed to get the bipolaron across in my lectures, the latest example of being stretched and found wanting. It will require another generation of textbooks to get the balance right.

One last picture. At the Condensed Matter and Materials Physics (CMMP) conference in 1995 Nevill attended for a day to present the Mott prize to Richard Friend, himself now Cavendish Professor. The CMMP series of conferences are direct descendants of the Manchester conferences founded by Nevill, traditionally held just before Christmas. Nevill made a great effort, travelled alone in cold weather by train, and needed occasional support to walk. I had to help him up the platform to present the prize; however, the stimulation of the event energised him; after the presentation he rejected my support and strode back to his seat with a firm unaided stride, picked up his newspaper and sat down. The following lecture described low temperature experiments intended to test the fundamentals of quantum theory; it very successfully engaged the attention of the audience of eight hundred or so, but it is not a topic which appealed to Nevill, who after his pioneering struggles with particle–wave duality in the early days of the theory, regarded it as settled and there to be used. He sat prominently at the centre of the front row, ostentatiously reading his paper. Nevill knew a trick or two about communication!

PROFESSOR L. M. BROWN, FRS, obtained his PhD at Birmingham in physical metallurgy and joined the Cavendish Laboratory as a postdoc in 1960. He stayed at Cambridge, eventually became a lecturer and professor in the Department of Physics. His main research interests are in the mechanical properties of materials and electron microscopy.

Moving On

Archie Howie

An essential component of Nevill Mott's remarkable ability to create and conquer successive fields was his decisiveness in cutting loose from the scene of his most recent triumph to free his mind for the next one. I joined Peter Hirsch and Mike Whelan just after they had seen the first dislocations in the electron microscope and given Nevill one of his most satisfying moments as Head of the Cavendish. For a few more years he remained active in the field and would occasionally summon me for a discussion about the scattering of electrons by stacking faults or to produce one or two illustrations for a review article on dislocations. I can still recall the sight of these impressive pages of handwritten text with scarcely a word crossed out. Gradually, however, he realised I think that the subject had lurched into a strongly experimental phase with clear-cut results offering less scope for his imagination. The scene was set for his transfer to amorphous materials.

Nevertheless, one more request unexpectedly arrived – for a picture of electron diffraction rings to take pride of place on the cover of the new edition of *Wave Mechanics*, one of the few books Nevill wrote without the benefit of a junior author who could keep it up to date after his interests had moved elsewhere. I made the mistake of trying to interest him in what was more readily available, a real-space image of electron interference effects such as thickness fringes or extinction contours. I received the legendary glassy stare. No words were exchanged but I was left in no doubt that anything different than a ring pattern would be a serious distraction in his task of getting the new edition out with the least possible fuss.

About 1970 I had a chance to experience Mott's equally decisive and widely underrated social deftness, when there was a visit from Jack Mitchell, his former Bristol colleague in the physics of the photographic process. When Nevill discovered that I knew Mitchell, he quickly communicated the idea that Churchill College would be a much more suitable venue than Caius for dinner on a cold Saturday evening in December. My college was somewhat surprised to find such illustrious guests dining that night, and although the Master was unable to join us, it was clear that the fiery cross had gone round and quite a large company was assembled, including our Senior Tutor. Nevill was in excellent form, telling us all about his first visit to Russia in the early thirties. Suddenly, however, at about 9 P.M., he stood up flexing his braces and saying that perhaps it was time to go, whereupon everything closed down very swiftly. I then found myself in the cloakroom with my two guests helping the senior one into an inordinately heavy overcoat. After a few seconds I became aware that Nevill had gone into reverse. He turned and said, 'Is there somewhere for Jack and I to have a talk?' adding, when he caught sight of my slightly puzzled expression, 'I thought that it was a good idea to let your Senior Tutor get off home.' Deciding that I myself was probably included in the group who should go off home, I showed Jack and Nevill back to the SCR which had been so expertly emptied by this stratagem and left them to it. Ever since I have taken with a very large grain of salt the numerous stories about Nevill Mott's lack of social skills.

Some (80th Birthday) Jots for Mott

New fields explore –
Mott's gift
For four score years
Festschrift!

Growth laws define
Oxides;
His mental shine
Abides!

Pick spins in flight
Mott dares,
Left-left, left-right
Compares.

Alloy that flows
Lengthens
And, as Mott shows,
Strengthens.

Photos retain
Each dot
Of silver grain
Said Mott.

Darkly we saw,
Pre-Mott,
Clear glass now law –
His plot!

What may conduct,
What not;
Rights to instruct
Mott's got.

Davis, Massey,
Sneddon,
Jones and Gurney –
Write on!

One tale that's far
Rehearsed –
Wife, Mott, house, car –
Dispersed!

Raise Gonville's glass,
Caius' pot,
Our toast lips pass –
Mott's tot!

A. H.

PROFESSOR A. HOWIE, FRS, has enjoyed an unbroken career in Cambridge. He began as a research student in 1957 when he was attracted from Caltech after reading a short Nature *review which Nevill Mott wrote about measuring Fermi surfaces. On arrival, however, he decided to follow Mott's alternative suggestion of research in electron microscopy and microstructural physics. This has been his main field of activity ever since, apart from the interruption of the last eight years, when he has been Head of the Cavendish Laboratory.*

Cavendish Professor

Brian Pippard

I was a boarder at Clifton College, in School House, like Nevill Mott fifteen years before. The scientists among us knew of his distinction, and when at only thirty years of age he was elected to the Royal Society, we were impressed and delighted. To judge from his autobiography, he did not enjoy his schooldays –

Four Cavendish Professors of Experimental Physics (1994): the order of succession is Nevill Mott, Brian Pippard, Sam Edwards and Richard Friend (proceeding anticlockwise).

to be clever without a love of games was no recipe for popular success in those days, and perhaps not even now. But he particularly praises his mathematics master, H. C. Beaven, and that I can heartily applaud, remembering Fuzzy B as a true mathematician and a most endearing man.

My first real contact with Nevill was when Sir Lawrence Bragg called me up from the laboratory to tell Professor Mott about my work. During the ten minutes of my little talk, he sat half turned away with his eyes fixed on the far corner of the ceiling; at the end he asked abruptly, 'Have you ever thought about bismuth?' The irrelevancy of the question was disconcerting, and I could make no adequate response; but a month or two later he was again in Cambridge and rang to say, 'I was interested in what you said about the anomalous skin effect; can you spare the time to talk more about it?' That was when I discovered his remarkable gift for storing what he heard while thinking about other matters. In middle age this ability began to fade, until like the rest of us, he had to concentrate on the matter in hand if he was to make progress. But all who knew him as a young man were astonished at his mental versatility, his enthusiasm for hearing all the latest ideas, and his apparently spontaneous comments that went to the heart of other people's problems.

At the time of Nevill's election to the Cavendish chair, the department was notably weak on theory, a legacy of Rutherford's time, when the emphasis had been on experimental nuclear physics. The ancient tradition of a mathematics faculty encompassing both pure and applied mathematics had concentrated the theoretical physics strength there, with fundamental particles and astrophysics as its principal foci; theoretical solid-state physics was not then, nor is it now, a serious concern of Cambridge mathematicians. Rutherford's son-in-law, Ralph Fowler, was poised between the mathematics and physics departments, and his successor, Douglas Hartree, though fully established in physics, had become a pioneer in electronic computing. G. I. Taylor, an outstanding member of the Cavendish and a powerful mathematician when he was not devising the kitchen-sink experiments which yielded him such penetrating insights, had his own fluid dynamics group which made little contact with anyone else in the department. So the solid-state experimenters, mainly low-temperature physicists and crystallographers, had little theoretical backing until Nevill's arrival. He was disconcerted to find no staff vacancies to enable him to introduce congenial new talent, but he persuaded the General Board to grant him a lectureship for John Ziman from Oxford. John was soon joined by Volker Heine, and from this beginning there grew a powerful team which still flourishes and covers a wide range of topics in condensed matter theory.

In the early years, however, research was by no means Nevill's sole concern; teaching and accommodation both demanded attention. The arrangements for the Natural Sciences Tripos were still stuck in the old pattern – two years devoted to studying three or four sciences chosen from an extensive menu (mathematics to anatomy) and only the final year available for a single subject like physics. The first change in 1958, for which he was an active campaigner, introduced advanced physics and chemistry as second-year options, but this was only a preliminary to a radical revision (1963) in which he was now the driving force. There was still a first year of choice among a more limited and better organised set of courses, which no longer included medical sciences, and this was followed by two years of specialisation. The pattern has survived except that the full physics degree now takes four years, in which the last allows the student to concentrate on a limited number of topics.

Student numbers in physics grew steadily during this period, while development of the site occupied by physics and other sciences gradually eroded the space available for practical classes. The number of research students also increased until no more could be accommodated, and some of the laboratories were in a state of advanced dilapidation. The whole of Nevill's tenure of the chair was punctuated by worries about buildings, and negotiations about the use of space occasionally caused rather severe frictions when a major research effort seemed to be threatened. Even before he had taken up the chair, in the interim after Bragg went to the Royal Institution, he was asked to decide the future of a half-designed linear accelerator which had rather reluctant Research Council support. Its abandonment was naturally hurtful to those involved, though with hindsight one can appreciate how it could never have been an effective rival of the huge machines that were already beginning to dominate the study of fundamental particles.

Not long after, a different kind of problem demanded firm action, but now it was a case of undeniable success – Perutz and Kendrew with their analysis of protein structures, Crick and Watson with the double helix of DNA. The resulting growth of molecular biology, perhaps the most spectacular of the Cavendish's triumphs, was more than could be contained on the site. If, as I believe, Nevill's initial thought was simply to see the last of these alien presences in a physics department, he soon recognised their achievements and needs, and helped to find them new laboratories where their ideas and discoveries could thrive.

To return to atomic nuclei and fundamental particles, the aborted linear accelerator was only the first stage of running down the dominant theme of Rutherford's Cavendish. The cyclotron, the Van de Graaff and Cockcroft–Walton machines were sold or broken up, and the vast hall that housed the last of them was converted into a fine lecture theatre with two floors of practical classes above. What remained of particle physics relied on experimental facilities at CERN and elsewhere, the data being analysed in the Cavendish. Robert Frisch's inventiveness culminated in Sweepnik, which could follow and record the tracks of particles in bubble chamber photographs. It was as precise as any rival, and much cheaper, but the Research Council was reluctant to believe it was anything more than a crude home-made gadget, unworthy of financial support. Nevill invited them to a demonstration after which, still unconvinced, they were subjected to an explosive verbal onslaught by Ken Riley (still a young staff member) that carried the day. Sweepnik kept the Cavendish in the business and later, having been adopted for uses like map-reading, it was the seed for one of the first enterprises on Trinity College's new Science Park, itself the child of a committee chaired by Nevill.

It was Nevill again who helped Philip Bowden transfer his laboratory for studying rubbing and friction from Physical Chemistry to the Cavendish. The PCS group (the initials were stable, their significance volatile) never became fully integrated on the old site, though it served as an object lesson on the desperate needs of physics for new quarters. When the department moved to West Cambridge, after Nevill's retirement, PCS was intentionally located at a major intersection of passageways so that it could no longer be ignored. As a result, it is now a leading member of the solid-state groups in the Mott Building, and has given us in Richard Friend the latest Cavendish Professor. All this would have gratified Philip Bowden, who was one of Nevill's staunchest supporters during the difficult times at Gonville and Caius College.

There were other building problems that helped to direct Nevill's thoughts to the needs of science departments at Cambridge. At first he agreed with the general belief that contact between researchers and the need for easy student movement between lectures and practical classes demanded concentration on central sites. But, as he soon appreciated, to rebuild a department means decanting its inhabitants elsewhere, and there was too little space for that. There was a serious plan to build towers of fifteen or more storeys, but strong objections were raised, which were shared by the inspector at a public inquiry. This is very fortunate because they would have allowed little opportunity for the chance meetings, as researchers go about their business, which can prove so fertile in generating new ideas. By the time it was agreed that a new site was needed for the Cavendish, Nevill had decided to leave the matter in other hands (including mine) while he devoted himself to the electronic behaviour of disordered materials, a problem he found fascinating and which paved the way to a satisfying and influential retirement.

If I have written about administration to the neglect of research, there is good reason. Our immediate interests rarely coincided and, however enjoyable our discussions, there was little I could say about his problems that he had not already thought of. And he did not wish, I felt, to be deflected from them to waste thought on mine. Others have written from intimate knowledge of this side of his life, but few had the opportunity to see him in action as a laboratory manager. His ways were his own, and they succeeded. In success he remained a man of great charm and kindness, whose eccentricities were lovable and whose friendship was inspiring.

SIR BRIAN PIPPARD, FRS, was educated at Cambridge University and, apart from four years on radar development during the Second World War, he has spent almost all his working life in Cambridge. His special research interests are in the electronic properties of metals, superconducting or otherwise. He has written several monographs and textbooks. He was Cavendish Professor from 1971 to 1982.

A Jewish Connection

David Tabor

In writing these reminiscences and appreciations of Nevill Mott, I may well reveal more about myself than about Nevill. I am aware of this possibility and I apologise in advance.

Nevill came into my life when he was appointed Cavendish Professor in 1954. I had returned to Cambridge with Philip Bowden at the end of the Second World War as part of his group. Philip had been appointed to a Readership in Physical Chemistry, although his research activities were not particularly representative of that department. A major change occurred in 1954 as a result of interdepartmental discussions and, I guess, as a result of Nevill's influence. Philip Bowden resigned from Physical Chemistry and was appointed Reader in Physics, his group becoming a sub-department of the Cavendish Laboratory. It was called PCS (Physics and Chemistry of Solids) and, despite many changes and fissions, it still retains that title. Our sub-departmental status in the Cavendish was a source of great pride and we went through a period of considerable excitement. Nevill's name became a household word and indeed was often invoked by our children. This had one unexpected consequence.

In traditional Jewish homes it is usual to chant the Sabbath grace after meals to traditional tunes and all members of the family join in. One of the phrases in the Sabbath liturgy refers to the Lord of Salvation and the Lord of Consolation. The Hebrew for consolation is *Ne-cha-mot*. At that time our younger son was breaking his teeth over the Hebrew and (perhaps in a playful spirit) found it easier and more homely to replace *Ne-cha-mot* by *Nevill Mott*. As a result, for a year or more Nevill's name was incorporated into our family grace and nobly ensconced with the Lord of Salvation. Although this episode occurred almost forty years ago, for my wife and me it still evokes resonances.

I got to know Nevill more personally after he resigned from the Mastership of Gonville and Caius in 1966 and came back to the Cavendish. We would often meet in the Cavendish cafeteria and talk briefly about social, political and ethical problems. Nevill had spent several months in 1928 in Copenhagen, one of the main centres of the New Physics, and in the following year he visited Göttingen, the other main centre. He had come back to Britain, inspired by quantum and wave mechanics, but he had also acquired first-hand contact with Jewish or half-Jewish physicists active in Copenhagen and Göttingen. In the year 1933, when Nevill was appointed Professor of Theoretical Physics at Bristol, Hitler was already active in the German parliament (the

163

Reichstag) and by 1936 his antisemitic Fascist regime was in full power. Nevill joined forces with other distinguished scientists and scholars to find openings for refugee scientists who could still escape from Europe. This was not easy because of unemployment at that time among British academics, and in some professional quarters there was opposition to the awarding of posts to foreigners. Nevill retained throughout his life his strong moral, intellectual and political support for the rights of minorities, his outspoken condemnation of apartheid and his sympathy for Jewish causes. I found I could talk freely to him about these and similar themes and recall especially my worries during the 1970s.

In 1973 Israel had been almost cut in two by Syrian tanks during the Yom Kippur War, and in the Soviet Union under Brezhnev, Jews who had been refused permission to leave the country (refuseniks) were singled out for constant harassment and often imprisonment. These issues unexpectedly found expression at a dinner held in Caius in 1975 to celebrate Nevill's seventieth birthday. Several short sparkling talks were delivered on his outstanding creativity as a scientist and as a man around whom many endearing anecdotes had collected. As the talks progressed, I grew rather restive and felt that something ought to be said about another side of his character. I asked the chairman, Abe Yoffe, if I could say a few words even though I was not on the list of speakers. He acceded. I spoke spontaneously from my heart rather than my head and had no notes. However, the Friedels were at the dinner and Mary Friedel, Ruth Mott's sister, asked all the speakers to write down what they had said for the family record. As a result I have a copy of my talk.

Among other things, I said, 'On occasions when I have been worried about the possible liquidation of the State of Israel, or the persecution of Russian Jewish dissidents, I have gone to speak to Nevill. I always found in him sympathy, understanding and goodwill. But perhaps more than this, I have always found a sort of rock-like moral strength which carries with it the conviction that somehow right and decency will ultimately prevail. It seems to me that these human qualities – his compassion, his humanity and his spiritual integrity – are as important and are as much appreciated by his friends as his outstanding attributes as a physicist.'

After the meeting, one of my friends said that my talk was okay, but wondered why I had raised moral and political issues at a function which was intended to celebrate Nevill's science. I wondered about this, too. But six months later, at a college dinner where farewell speeches were being made to a colleague, Nevill happened to be sitting opposite me. He leant over and said, 'I shall never forget what you said at my seventieth birthday.' This was not only balm to me, it made me realise how deeply Nevill felt about issues that rarely received public expression.

In one of our brief conversations I raised with him his outspoken opposition to the suppression of Black South Africans by the ruling classes; I asked why he had not been pro-Arab in the Middle East dispute and why he had been pro-Israeli. He answered in five words, 'You must make your choice.'

This was not a snap decision but the condensed summary of very deep thought, as I discovered during a group discussion when he gave a more detailed account of his thoughts. He had clearly made his choice.

Nevill began to be seriously interested in religion in his fifties, and when he turned 80 he expressed his commitment by accepting baptism. He felt the need for more open discussion by scientists about their religious beliefs. He approached scientists whom he knew and asked them to contribute a chapter on their views to a book he was editing entitled *Can Scientists Believe?* When it was published in 1991 Nevill wondered why the contributors were so keen on dogma and ritual. His own chapter deals with Christianity almost entirely in terms of ethics and morals; he accepts Jesus's message but little of Christian doctrine. His approach is that of natural theology. As John McDonald (Chaplain at Caius) wrote in his obituary notice in *Church Times*, it is not surprising that the Church Establishment did not use him as they could have done and as he would have wished.

Most of these reminiscences have a Jewish connection. My last item deals with a far more general theme. It also reflects one of Nevill's striking qualities – he could hear what was said to him and digest it without apparently listening. About a dozen years ago, a Cambridge professor who had been head of his department, head of his college and a 'short-term' vice-chancellor, suddenly died in his early sixties and I attended his memorial service. When I came out, I found myself standing next to Nevill and I wondered if any words were necessary or seemly.

Nevill was a man of grand silences. But I am not, and I said, 'Sad business.' Nevill nodded. Then I said, 'It's a great pity he died so young.' Nevill nodded. Then finally I added, 'I suppose it is simply one of the hazards of being alive.' Nevill nodded. Then suddenly his whole expression changed, his face lit up, his eyes sparkled and he said, 'But David, it's worth it.' These few words summarise in typical succinctness his attitude to life, to work and all else.

PROFESSOR D. TABOR, FRS, was born in London in 1913 and educated at Imperial College, London, and at Cambridge, where he enjoyed over thirty years of collaboration with Philip Bowden. His main research interests are friction, lubrication, hardness, surface forces, colloids and polymers. He is now Emeritus Professor of Physics at the Cavendish and a Fellow of Gonville and Caius College.

NFM and PCS

Abe Yoffe

It was with some relief that I heard from Philip Bowden in 1956 that the Physics and Chemistry of Solids Laboratory (PCS), to which I then belonged, was to join the Cavendish as a sub-department, rather than move to the new Physical Chemistry Laboratories in Lensfield Road. I had returned to Cambridge from the Weizmann Institute in Israel in 1954 as an ICI Fellow, and my interests then also included the electronic properties of the metal azides, and so my bible was the book by Mott and Gurney. Nevill then had an interest in the silver halides among the many other topics he was working on, and there are indeed many similarities in the behaviour between these two classes of materials. At that time, Nevill appeared to me as a tall, rather aloof and detached man, who liked to talk physics and chemistry, but who rather hated small talk. This soon became apparent to anyone attempting this by his

Abe Yoffe's retirement in 1987: Abe listens to Nevill and Ted Davis is in the centre.

repeated glances at his wristwatch. He was to have a great influence on both my research activities and my career, and without his support then and in the later years, I would not have been able to remain in Cambridge. I did not collaborate directly with him on any specific topic, since this would not have worked out, given our approaches. But through contact with him, I moved into areas such as semiconductors, low-dimensional solids, amorphous solids and metal–insulator transitions. He attracted numerous scientists from around the world, who came to work with him or to have discussions; I was able to meet many of them and several have remained as colleagues and friends to this day.

As Master of Caius College, Nevill joined with the Masters of Trinity and Johns as the main trustees who helped found Darwin College in 1964, and I was appointed with eleven others as one of the founding Fellows, together with the first Master, Sir Frank Young. I suspect that Nevill and Philip Bowden were partly involved in my receiving the invitation to help set up Darwin as the first graduate college in Cambridge, and also as the first mixed-sex college. He took pleasure in seeing it flourish and grow to its present size of almost four hundred graduate research students, and in 1978 he was elected to an Honorary Fellowship in the college.

Nevill retired early from his Cavendish chair in 1971, and since he already had many ties to PCS, where he felt 'at home', it was no surprise that he came to us for a room. The other incentive was that we occupied space facing Free School Lane, which I guess was about as far removed as you could get from the 'seat of power' in the Austin Wing of the department. Staying on after retirement was something of a novelty then, but as far as I remember, there was no objection to this and in a way it set a precedent for the subsequent elderly retirees in the Physics Department. In 1971 there were already plans to move to West Cambridge to a site being masterminded by Brian Pippard, and Nevill welcomed with enthusiasm the suggestion that he should come with us and 'live' among our experimental group. He had always made clear his wish to mix with experimentalists, theoreticians and scientists from different disciplines when their interests overlapped. Ted Davis will doubtless elaborate on this when he writes of their joint work on amorphous solids.* In later years both Ted and Christine were to become very much like a son and daughter to him. It was also characteristic of the man that he went out of his way to spend time and effort in order to make our overseas and continental students feel welcome in the lab. Nevill stayed with us for over twenty-five years, much to our benefit and pleasure, and when I retired in 1987, Yao Liang then encouraged him to join him in the IRC on Superconductivity, a topic of interest to him until the end.

In the early eighties, Nevill wrote his autobiography and gave me his handwritten chapters to read as they reeled of his production line in very quick succession. I did in fact make some suggestions for changes; he used to

* See the introduction.

thank me nicely, but took little notice, apart from agreeing to include a number of photographs. Given his great self-belief, this result is perhaps not surprising. At this time he also seemed to involve himself in religious questions and showed me the contents of his talks before delivering them in churches, but again he was not inclined to follow many of my suggestions.

By then he also felt he had done as much as he wanted on amorphous solids, and frequently discussed with me possible new areas in solid-state physics that should be tackled. I remember trying to sell the idea of looking into the electronic properties of organic semiconducting solids, and he did in fact obtain a copy of that excellent book by Pope and Swenberg on electronic processes in organic crystals. However, after looking through it, he felt he did not want to start on a course in organic chemistry and decided the topic was not for him; he gave me his copy of the book. Nevertheless, he approved very strongly of the work now being carried out on the optoelectronic properties of organic polymers by Richard Friend, the current Cavendish Professor. As for the PCS lab that he knew very well, he mused at a lunch to celebrate his ninetieth birthday, that bringing it into the Cavendish, back in the fifties, was possibly one of the best things he did as Cavendish Professor.

For me, like many others who knew him well, his generosity, his humanity, and above all, his powerful intellect, which remained with him to the end, these are the things I shall continue to cherish. He was a great, good and kind man.

DR A. YOFFE was born in Jerusalem in 1919 and obtained BSc and MSc degrees at the University of Melbourne in physical chemistry. After war service in Australia he came to England in 1945 and joined Philip Bowden in the Department of Physical Chemistry at Cambridge. His research group, Physics and Chemistry of Solids (PCS), eventually became part of the Cavendish Laboratory. Dr Yoffe has wide interests in the field of condensed matter physics.

Gifts to Physics

John Ziman

What did Nevill Mott give us, as fellow scientists? I don't mean his actual contributions to physics, the particular items in the long catalogue of imaginative concepts, intuitive inferences and theoretical explanations that made him so creative and famous. Many of these contributions are still holding up their heads in the torrent of scientific progress; others have served their turn and passed into the historical record. What they happen to mean now is almost irrelevant to their eternal scientific significance, which was what they meant when they were newborn, twenty, forty, even sixty years ago. Even for those of us who can recall our delight when we originally received them, it would require an elaborate exercise of memory and mental archaeology to reconstruct the scientific backdrop against which they stood out so vividly.

What this exercise would certainly reveal is how brave, bold and unexpected these contributions were. We saw them at the time as the products of an audacious intuition, leaping to startling theoretical conclusions on relatively sparse and often confused experimental evidence. It was not easy for others to adopt his methods of thought and argument; in fact, it would have been unwise without Mott's peculiarly broad grasp of physical phenomena and mathematical principles. But even when his novel ideas did not seem immediately convincing, they were always so challenging and so clearly formulated that they provoked positive experimental research or theoretical effort to confirm or confute them. Once again, it would take a very detailed analysis to draw up a balance sheet of the eventual yeas and nays. But for more than half a century they were an unfailing source of intellectual stimulus for the whole field of what we now call condensed matter physics.

In a wider perspective, Mott's gift to our subject was a set of keys to the gateways into a succession of new fields. They were often put into our hands through the books that he coauthored, reviewing and analysing whole classes of systems and phenomena which had previously been disregarded, showing how these might be interpreted and understood. By the publication of yet another 'Mott & X', volume, and by other strategic initiatives such as informal conferences, invitations to carefully chosen visitors, choice of thesis topics, etc., he would shepherd us over the threshold into this new domain of enquiry, handing out a preliminary agenda for research as we entered.

I say 'gifts', for Nevill asked very little in return for what he gave. The most was a request to be told what one was doing, thinking, discovering, explaining or not understanding. That could sometimes be disconcerting. He would stand quietly looking out of the window, occasionally puffing at a cigarette or taking off the jumper that he had put on ten minutes before, but Buddha-like in not showing any outward sign, such as a nod or a grunt of assent, that he was taking in your increasingly insecure discourse. But the eventual reward for going on was a shy but dazzling smile, and a gentle question into the heart of the matter. Unlike many great men, he was a good listener, and a lifelong seeker of information and ideas that might throw light on the difficult problems with which he was grappling. He was also very generous with his own thoughts as they developed, although I have to admit that his line of argument was not always easy to follow until it had crystallised and matured.

For the ten years I was at the Cavendish, we often discussed Cambridge ways and means. Coming from Bristol, he saw at once that the Natural Sciences Tripos was educationally clumsy and very ill-suited for the combination of mathematics and physics required for research in theoretical physics. This was the beginning of a long, slow battle for reform, which he did not completely win – a gift that physics has still not yet received. But the Cavendish was a happy and very productive research laboratory and teaching department under his leadership, just as Bristol physics had been in his years there.

Perhaps it was simple friendship that was his richest gift. In spite of a formidable personal and scientific presence, Nevill Mott was very obviously a gentle, humorous, tolerant man. I recall Hans Bethe teasing him with the old chestnut about the absent-minded professor who has to ask, at the end of a chance conversation, whether he was coming in or going out of the building, in order to decide whether or not he has had his lunch. 'Hans,' said Nevill, 'I've had that story told about me in three different languages, and it's not true in any of them.'

I don't know what everybody else in this grand collection of friends will be saying, but I guess they will all mention the kindness that he showed to the numerous younger people whose company he so obviously valued. I was one of that number, and am very happy to express my gratitude for just such kindness over many years as a colleague and friend.

JOHN ZIMAN took his DPhil in theoretical solid-state physics at Oxford then accepted a lectureship at Cambridge in 1954, nominated by Nevill around the time he became Cavendish Professor. Ziman went on to a chair of theoretical physics at Bristol in 1964, where he eventually succeeded to Mott's old post, Director of the H. H. Wills Physics Laboratory. He retired as Emeritus Professor in 1982.

A Reminiscence and Appreciation

Alan Cottrell

Nevill Mott has for so long been the great father figure of solid-state physics that we tend to overlook his earlier contributions to fundamental quantum mechanics. But his analysis which solved the mystery of how a spherical wave function, spreading outwards in all directions, collapses upon measurement into the sharp, straight track of a small particle, continues to be quoted by those involved with the foundations of the subject.

What led him from this into solid-state physics? He has given a few clues himself, and others will know the story better than I. But I have a theory. Anticipating in 1933 his many future decades of intense research, he wanted a field so rich in research problems as to be virtually inexhaustible. And what better for this than a solid, with its 10^{24} atoms and electrons all a-thrill? Thus, armed with the ladder of quantum mechanics, he climbed the wall into what was then a secret orchard.

It was vast beyond belief. Enough not only to satisfy him through all those decades, but also to enable him to realise a second ambition, that of encouraging us all, now in thousands, to join him in enjoying its riches. This he did both through his many introductory books and also through the delightful conferences and summer schools which he organised, those at Bristol in the early postwar years being particularly outstanding. They were like grand orchestral concerts at which, under his leadership, we all played far above our normal best. Of course, it did not always go faultlessly. I recall an occasion when, after a session on rheology in which the concept of a Bingham body was discussed, following which we were enjoying drinks, he espied a solitary, dignified individual standing aside. Going up to him, Nevill said cheerfully, 'Ah, you are the Bingham whose body we have been studying.' 'My name is Bangham, not Bingham,' came the daunting reply.

His delight in spreading the gospel is obvious in his books, where even at their most advanced, the subject is presented in the simplest and physically most intuitive way, never going beyond the bare minimum of mathematics. This approach occasionally attracted criticism from the formalists. But what always mattered for Nevill was the physical picture of what was going on. Mathematics was merely a tool, to shape this picture. He developed this almost back-of-the-envelope approach to such an art that he reckoned, when embarking on a new problem, that it was quicker to sit at his desk and create a preliminary theory than to go searching through the vast literature to see what had gone before.

Newton is said to have made his discoveries by 'thinking at all times on these things.' It was much the same with Nevill; and many stories spring from this. John Gilman vividly recalls seeing Nevill and me, in 1956, standing knee-deep in the freezing waters of Lake Placid, so lost in scientific discussion as to be oblivious of our surroundings. Evidently his concentration equally claimed my attention on that occasion.

Even driving his car did not break his scientific concentration, for his style as a motorist was, shall we say, semi-detached. My wife still blanches at the recollection of how he used to back his car out of our drive and into the main road without regard to the passing traffic. His safe emergence from many years of driving must be counted as the strongest possible evidence for the existence of guardian angels.

Another example: at a time when I was at the Atomic Energy Research Establishment (AERE), Harwell, Nevill paid it a visit. Several of us there took him to lunch. The conversation drifted on to Winston Churchill, who had also just visited Harwell, and his role in the war. Nevill said nothing but followed the conversation round the table and occasionally gravely nodded assent. Eventually the conversation petered out. He immediately brightened and spoke at last, 'I think I have it. Suppose that the magnetic field is along the hexagonal axis and that the electrons are moving in the basal planes.'

Although his attention was not easily drawn away from science to more worldly matters, once engaged it could be sharp, firm and sometimes devastating. I recall an occasion when a university chair was being advertised. I showed some interest in it, mainly because it was situated in a particularly attractive, if remote, part of the country. His reply, 'Well Alan, if you really want to bury yourself out there ...' immediately killed the idea stone dead.

He had a clear understanding of the academic hierarchy in Cambridge, expressed in his words, 'Professors are the lowest of the low. The only thing to be is the head of a college.' Thus my wife and I were not surprised at his delight when he called on us, one day, to say that he had just been elected to such a headship. I think that nothing had pleased him more, and so his disappointment was all the sharper when, a few years later, things developed acrimoniously in his college.

Nevill gave little thought to issues of national scientific policy. Quite early he had concluded that efforts in that direction were a waste of time: 'Politicians never listen to scientists. Better to put your energies into something else.' And he followed his own advice, with excellent results; for example, his pioneering venture led to the setting up of the Cambridge Science Park.

How shall we remember him? His science? Textbooks and research papers will take care of that, even those yet to be written in the coming millennium. But personally? We all have our own memories. For me, one of the best is of those Bristol conferences with Nevill in command, his voice specially turned up for the occasion – and he could raise its volume immensely in full, cheerful measure – ringing out across an enchanted audience.

SIR ALAN COTTRELL is a Distinguished Research Fellow in the Department of Materials Science and Metallurgy, University of Cambridge. His previous appointments include Goldsmiths' Professorship of Metallurgy, University of Cambridge, Master of Jesus College and Vice-Chancellor, University of Cambridge, and Chief Scientific Adviser to Her Majesty's Government.

Applied Science

James Menter

Nobel laureate Nevill Mott has been quite properly described as one of the most distinguished theoretical physicists of this century. It is all the more remarkable that he maintained an abiding commitment to the application of science to useful ends throughout his working life. The urge to find uses for his prodigious intellectual talent outwith the narrow confines of the academic world is a continuing theme permeating his fascinating autobiography, *A Life in Science*. It is evident in the pioneering work he did before the war in Bristol, which gave him his first intimate contact with industry.

It was through my employment as an industrial scientist that I came to know him personally after he was appointed Cavendish Professor of Experimental Physics at Cambridge in 1954. In that same year I left F. P. Bowden's laboratory in the university to join an industrial research laboratory (TIRL) newly established at Hinxton Hall, some ten miles south of Cambridge, by Tube Investments Ltd, a large engineering manufacturer. Bowden was an intimate friend of Mott and both encouraged me to maintain a promising research relationship with P. B. Hirsch and his colleagues in the Cavendish, a relationship which had developed before I left Bowden's laboratory (at that time located in the Physical Chemistry Department). D. W. Pashley also joined TIRL a few months later. The Hinxton and Cavendish groups each acquired a high resolution electron microscope, with which they were able to open up new vistas in the microstructure of crystalline solids. The common scientific interests of the two laboratories created a lively traffic in ideas and people between them.

The following entry is recorded in the laboratory diary of P. Duncumb, a research student who invented the scanning electron probe X-ray micro-analyser in V. E. Cosslett's research group in the Cavendish: '7 August, 1957. Worked with D. A. Melford of TI on segregation of copper and nickel in steel. Finished in one day.'

It turned out to have been a very productive day's work on a practical problem that had arisen in a TI steelworks. Duncumb and Melford collaborated to design, construct and install at Hinxton a fully engineered version of Duncumb's Cavendish apparatus. This was the prototype for the subsequent manufacture and sale under licence of some eighty instruments by the Cambridge Instrument Company, which thereby became a major player in the international market for advanced electron optical instruments.

These are but two of what must be many such interactions between the Cavendish and industry during Mott's tenure. They serve to sketch in the background, in his own laboratory, to his seminal role in chairing the committee set up later in the 1960s to formulate the university's attitude to local science-based industry. In his own words: 'The *Mott Report* (of this committee) was instrumental in enabling Trinity College to open its Science Park, to which the Master of that College paid generous tribute in the opening ceremony in 1975. The massive development of technological industry in Cambridge is now (1985) referred to as "the Cambridge phenomenon".'

Nevill Mott was wonderfully far-sighted. In science he was a brilliant strategist, choosing problems with far-reaching and beneficial consequences and in the world of affairs his vision and leadership contributed enormously to improving relationships between academia and industry to mutual advantage.

At a personal level I always enjoyed his company. I confess that, at first, I thought him rather forbidding until I realised that he was not given to an ill-considered response to one's remark. This slightly embarrassing silence while one went on talking nervously would be broken by a profound comment on a point several sentences back in the conversation, as likely as not accompanied by that disarming smile.

The last time I spoke with him was at the gathering in Oxford in honour of Peter Hirsch's retirement in 1992, when I was delighted to find him as my neighbour at dinner. I listened in awe as he spoke with such enthusiasm of his continuing research in superconductivity. I mentioned that I had read about the appointment of a Russian scientist from the former Soviet Union to an N. F. Mott Fellowship and said it was a splendid thing that this award bore his name. His response was typically direct and to the point, 'Oh, I organised that, as I needed someone to talk to about superconductivity.' His astonishing scientific productivity continued through his eighties and the Christmas cards he sent me bore some characteristically succinct message indicating that he was still in business. The last one came in 1995; it simply said, 'Still doing physics, high T_c superconductivity.'

The most endearing letter I ever received from Nevill was his reply to my message of congratulations when he was made a Companion of Honour. It was not the sort of letter to which one would necessarily expect a reply, but some weeks later an envelope arrived at my home in Scotland, addressed to The Occupier. And inside was not some boring letter from a bureaucrat but a handwritten letter from Nevill, starting 'Dear Jim, I couldn't remember your surname.' His memory may have been failing him, but he remained undiminished in his remarkable capacity for problem solving on the basis of limited experimental evidence, in this case my first name and the printed address on my notepaper. I treasure that letter – and the envelope!

SIR JAMES MENTER, FRS, served with the Admiralty during the war before going on to study physics at the University of Cambridge. He carried out his postgraduate and postdoctoral research in the Physics and Chemistry of Solids Laboratory (now incorporated within the Cavendish Laboratory). In 1954 he joined Tube Investments and became its Director of Research and Development in 1960. In 1976 he was appointed Principal of Queen Mary College, University of London. He has served as President of the Institute of Physics.

Plate 1   Nevill relaxing in the United States (1930s).

June 1938

H B David   W F Cox   J D Eshelby   W R Harper   W E Matthews   S R Tibbs   J R Bristow

N Arley   H London   N Thompson   B R A Nijboer   R L Mercer   G I Harris   H Frohlich   C R Burch   R W Gurney   J H Burrow   C F Powell

W Sucksmith   L C Jackson   I Williams   S H Piper   N F Mott   A M Tyndall   E T S Appleyard   W Heitler   H H Potter   H M O'Bryan   H W B Skinner

Plate 2   Academic staff of the Physics Department, University of Bristol (1938).

Plate 3 With Chinese visitors at Trinity College (1947): Nevill in characteristic pose during a meeting to celebrate the tercentenary of Newton's birth.

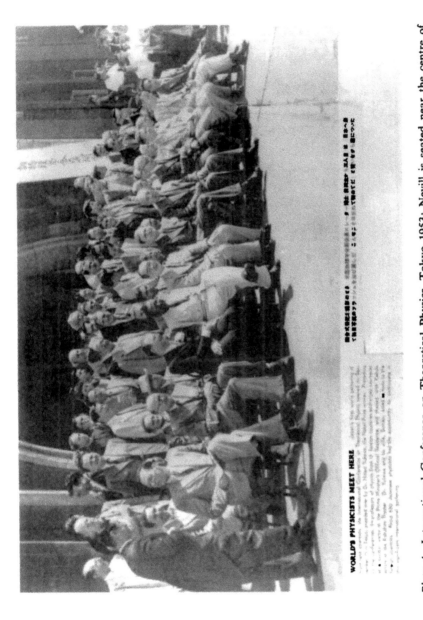

Plate 4 International Conference on Theoretical Physics, Tokyo 1953: Nevill is seated near the centre of the front row and on his left is Dr H. Yukawa, a Nobel prizewinner.

Plate 5   Nevill receiving the Nobel prize for physics from the King of Sweden (1977).

Plate 6 Celebrating Nevill's Nobel prize at the Cavendish Laboratory: the large bottle of champagne was a gift from Stan Ovshinsky.

Plate 7 Nevill speaks after planting a tree near Lake Balaton, Hungary, in celebration of his Nobel prize.

Plate 8   Nevill in front of the Mott Building at the Cavendish Laboratory.

Plate 9   Nevill with young scientists at a conference.

Plate 10   Pope John Paul II greets Nevill at the Vatican during a meeting for Nobel prizewinners.

Plate 11   Nevill in 1981.

Plate 12 Maggie Smith, the actress, chats to Nevill on the day they receive honorary degrees from the University of Cambridge (1994). (Courtesy Silver Studios, Huntingdon)

Education

Energetic Kindness

Martyn Berry

Other writers will cover the grand sweep of Nevill's work in physics and his immense contributions to education and public policy. Here I shall offer a personal response to probably the most remarkable man I have yet met.

In 1968 Nevill was chairman of a working party set up by the Royal Society and two other bodies to look into the shortage of maths and science teachers in secondary schools. I had written some pieces critical of government policy and spoken on the theme at the 1967 British Association annual meeting, and presumably this was the reason I had been asked to collect the data and do the fieldwork for the eventual report. (The conclusion was that a real problem existed. I also found that Sir Edward Appleton had produced a similar report fifteen years previously. The position is much the same now. Some things hardly change.)

After the publication of the report late in 1969 Nevill suggested that I might be able to help the Taylor & Francis subsidiary, Wykeham Publications, as an editor-cum-schoolteacher, a co-author for a series of books aimed at able A level pupils. He even enlisted me to 'help' with his *Elementary Quantum Mechanics*. I recall that I contributed one genuine idea, adjusted a couple of paragraphs to fit the intended readership, made a few suggestions to assist clarity, and that was about it. I confess that I learned more from Nevill's manuscript than I had from Charles Coulson's lectures at Oxford, probably because this time I was actively concentrating.

In the early and mid-1970s I served on various Royal Society committees, especially the Chemistry Education Committee, jointly arranged by the Royal Society and The Chemical Society. Nevill was chairman of the corresponding physics committee, and we met and corresponded from time to time. He had, and retained to the end of his life, a passionate concern for education at school level, especially for providing the widest possible opportunities so that all children could experience good science teaching. His reasons were stated in the introduction to the 1969 report: 'Such teaching is vital if the nation is to have adequately trained scientists and engineers. But ... in these days it is increasingly important for all educated people to have some understanding of mathematics and science.' I recall that his insistence on the absolute need for a scientifically literate democracy featured often in later discussions.

As chairman of the Physics Education Committee, Nevill was an ex officio member of the Royal Society Education Committee (RSEC). In April 1975 he wrote for advice:

I thought I would let you know the way our present Education Committee is going. Sir Harold Thompson, whom you quite certainly know, is taking a 'Black Paper' point of view about standards of science teaching in schools and wants to have an open meeting where Fellows get up and denounce the present situation and call for a return to the old ways. Our Chairman, Clifford Butler, as you would imagine, feels quite otherwise, and I gather that there was a meeting – which to my great regret I missed because of trouble on the railways – when there was a real slanging match between them.

Nevill invited me to a meal to discuss the situation. I accepted with pleasure. I had already heard from Malcolm Robinson, Education Officer of the Chemical Society, what had happened at that remarkable meeting; and although I had great respect for 'Tommy' Thompson as a pioneering spectroscopist and as President of the International Union of Pure and Applied Chemists (IUPAC), I had myself come up against his built-in prejudices and steam-rollering tactics more than once already. (Tommy may most clearly be lodged in public memory as the President of the Football Association who sacked Sir Alf Ramsey from managing the England team.) I like to think that I contributed some small part of the ammunition which enabled Nevill and the other 'moderates' eventually to win the day, ensuring that the goodwill earned by the RSEC from hard-pressed schools was not destroyed in one intemperate outburst.

We continued to correspond, and to meet once or twice a year when I took a small party of sixth-formers to the Cavendish, but more usually when Nevill invited my wife Jill, also a chemistry teacher, and myself to a meal in the Ladies, Annexe (marvellous anachronism) of the Athenaeum. It was a particular delight to attend his eightieth birthday celebration dinner at the Athenaeum. I think he enjoyed using us as a test bed for ideas, knowing that he would at least get honest responses, often imperfectly informed but uninfluenced by allegiance or obligation owed to any person or institution. After his Nobel prize in 1977 he told us in a totally unaffected way that he intended to devote most of the prize money to the education of his grandchildren. This contrasted with another Nobel laureate, who used his to buy a yacht.

Conversation increasingly centred around Nevill's growing interest in the links between science and theology, and in the morality of nuclear weapons. It took great effort to try to keep up with him in either field; he had read and thought deeply in both, and although he never openly challenged my own beliefs, he certainly caused me to think very hard about them. He asked me to contribute to his book *Can Scientists Believe?* I now regret very much that I havered too much about what I would write, and eventually backed out. Nevill was very patient and understanding.

Nevill was a firm advocate of 'No First Use' for nuclear weapons. He was active in Pugwash, and established fruitful connections with the Oxford Research Group, which studied the ways in which decisions were made about nuclear weapons. He once complained to us, very mildly, that his Nobel prize meant he was often being asked to lend his name to good causes, particularly in the fields of education and nuclear policy. He felt he had to be very careful and selective.

He was scathing about President Reagan's Star Wars initiative, and gave good reasons for his scepticism. He wrote, and asked me to comment on, a well-balanced booklet on nuclear weapons for the *Science in Society* series, designed for an A/O exam at sixth-form level. I asked him, rather diffidently, whether he would come down to Sidcup and talk about these matters to our Lower Sixth. To my pleasure and mild surprise he accepted; he did a very good job in stimulating a large audience of 16–17 year-old sceptics for whom great age and distinction meant little.

Encouraged by this, I then asked whether he would care to address the Sevenoaks Peace Forum on a similar topic in the summer of 1987. He agreed, and this led to the twelve hours or so which encapsulate the essence of what Nevill meant to us.

On the day, I collected Nevill from Sevenoaks station at about 6 P.M. and brought him back to our home in Seal. He talked with charm and vigour over a cup of tea before I took him to supper with the meeting's organisers. He ate heartily and continued to talk with animation until I whisked him off to the meeting itself. The title of his talk was 'Deterrence and the Pugwash meeting on Star Wars: some ethical considerations', apparently rather heavy! Certainly it lasted about forty minutes and was, frankly, rather stilted. He then demanded questions. For the next hour and a half he strode about, occasionally tripping over chairs, but all the while arguing, cajoling, gesticulating, striking sparks from a willing but largely sceptical audience, compellingly fluent, careful in listening and quick to respond. It was quite difficult to prise him and others away; the caretaker stood for some time jangling his keys before the meeting broke up.

Back at Seal he cheerfully talked about university life over cocoa and biscuits with our elder daughter. Well after midnight I managed to direct him to the stairs. Halfway up he stopped. 'Oh, I forgot to mention it,' he said. 'I have to go to a conference tomorrow. Could you get me to the station by seven?'

When I took him tea at 6 A.M., summer sunlight was already streaming through the open curtains. Nevill was sitting up in bed with somebody's PhD thesis on his knees and a pen in his hand. 'The argument's sound enough,' he said, 'but why on earth can't people write decent English these days?' This incident happened a couple of months before his eighty-second birthday.

We met Ruth only twice; the impression remains of a charming and very capable person to whom Nevill was obviously devoted. Looking through his letters now I find much about the family, and particularly about the grand-children. For all his myriad activities and concerns and what Tam Dalyell in his *Independent* obituary called Nevill's 'demonic energy', he clearly gave a great deal of time to his own family and to his friends. I think that he simply enjoyed people. It seems to me that one of the messages to come from his autobiography, *A Life in Science*, is that science is, of necessity, a people-centred activity. And I think that one of the reasons why, in the nearly thirty years that we knew him, our response to Nevill grew from respect to near reverence was that, so far as we were concerned, he was unique in his combination of intellect, humility, straightforwardness and totally unaffected kindness.

His last letter to us was written in April 1996, and re-reading it now moves me deeply. The only sign of advancing age is that he repeats some things which he had already told us. But many of the main themes are still there: concern with education, trenchant views on current affairs, delight in his work (here complete with a gently maverick glee), family news, interest in theology, and so on:

Congratulations on your struggle to maintain decent education. Today it looks – after Tamworth – certain that we shall have Tony Blair. I wish I had any confidence in him. . . .

My personal news – I had my 90th birthday in Sept, and also a CH from the Queen; a 10 minute interview and I found her charming.

But my personal life has been sad because my wife has Alzheimer's disease. ... I spend a few days in Cambridge most weeks and see her. [There follows news of all three grandchildren.]

In physics I have acquired a Russian colleague (Alexandrov) and together we have written two books on high T_c superconductivity; differing completely from Phil Anderson and most of the Americans. Keeps me alive! Also I remain very interested in science and theology. It seems to me certain that the Big Bang theory is right and time began with the Big Bang (did God exist before time?) The old dispute Einstein vs Bohr is live as ever. I am a Bohr man. I believe that the future is unknowable and God is not omniscient (or omnipotent).

I look out of my window and see some unwelcome snow on the lawn.

What a superb, questing brain. What a marvellously complete man.

MR M. BERRY read Chemistry at Oxford. He has taught at Chislehurst and Sidcup Grammar School since 1962, being Head of Chemistry 1966–1982 and Joint Head of Sixth Form since 1976. He was President of the Education Division of The Chemical Society 1973–1974, a member of the RS/CS Chemistry Education Committee 1968–1977, a member of the British National Committee for Chemistry 1973–1978, and received the 1976 Chemical Society Award for Chemical Education.

Starting a University

Roger Blin-Stoyle

Some thirty-five years ago I moved from the secure life of Oxford University to take part in the setting up of the new University of Sussex. This move was prompted by an exciting year spent teaching and researching at the Massachusetts Institute of Technology (1959–60) and the desire, on returning to Oxford, to change the structure and delivery of its physics courses. This was not possible on a short timescale and the alternative possibility of putting my ideas into operation at once in a new university was too attractive to be resisted. Of course, some preliminary ideas about the academic structure at Sussex and the nature of its courses had already been undertaken by an Academic Advisory Committee, consisting of a few very distinguished people, which had been set up three years before the university actually came into being (1961). Nevill Mott was a key member o this committee and had considerable influence in formulating its preliminary general ideas about the nature of the science courses.

These ideas were set out in a brief note about the Faculty of Science in the first (Arts and Social Studies based) prospectus and, as far as science courses were concerned, it stated that

> There will be a three-year course in Natural Science, leading to a classified honours degree. During this course students will study three subjects chosen from Pure Mathematics, Applied Mathematics, Physics, Chemistry, and Biology. In addition, the curriculum will require some study (which will be examined) of the history of science and the impact of science on society.

It went on:

> For a limited number of selected students there will be a second honours course lasting one year, in which one science only will be studied, and a successful candidate will be awarded classified honours in that subject.

This structure, which was not in fact implemented, is remarkable in that in many respects it anticipated what was to be initiated over three decades later with the advent of the fourth-year MPhys, MChem, MMath and MBiol courses. It was an imaginative proposal that came well ahead of its time but which, apart from not being financially viable, went against the strong feelings of those of us who staffed the university in its early days. We believed there should be specialisation within the basic three-year degree course. So, contrary to the ideas of Nevill Mott and other members of the Academic Advisory Committee, from the outset science courses required undergraduates, after a more general two-term preliminary course, to major in physics, chemistry or mathematics (biology did not start until 1965).

In conformity with the Academic Advisory Committee's view that the sciences and mathematics should be to some extent integrated, all science undergradates during the years 1962–65 belonged to a single administrative unit, the School of Physical Sciences, of which I was Dean. This had considerable benefits and worked well to begin with. However, it was clear that after a few years it would be too large (around eight hundred undergraduates) to be a unit which could operate cohesively and which could develop a sense of belonging in the minds and activities of undergraduates. It was feared that, for administrative and convenience reasons, the school would inevitably subdivide into a group of fairly autonomous departments (a forbidden, because divisive, word at Sussex). For this reason, in order to preserve the concept of a school, I proposed that the School of Physical Sciences should itself fission into two new schools, the School of Mathematical and Physical Sciences (mathematics, physics, philosophy of science) and the School of Molecular Sciences (mainly chemistry). This idea, which was approved within the university had also to be approved by the Academic Advisory Committee. All members agreed without argument except Nevill Mott! He wrote as follows to the Registrar:

> I am a little embarrassed by your letter. The decision you are making about your scientific schools seems to me of great importance and would be noted by the whole country.

and to me:

> I do feel that this is a very important decision not only for Sussex but for scientific education in the country, and I simply cannot say whether I approve of it without a detailed examination of the position.

As a result of these letters (both dated 22 January 1965) I went to Cambridge on 1 February and lunched with Nevill Mott at Gonville and Caius to discuss the proposal in detail. He was considerate, kind and understanding in this discussion and it was a great relief to me and all my colleagues that he finally agreed to the proposal; he wrote on 2 February to the Registrar confirming this. It is perhaps ironic that, in 1996, physics and chemistry came together again in a new school, the School of Chemistry, Physics and Environmental Sciences, and mathematics went its own way in the School of Mathematical Sciences!

Soon, as the university continued to grow, the Academic Advisory Committee was disbanded and Nevill Mott's formal influence and my direct contact with him ceased. However, our paths crossed again in the late 1970s and early 1980s, when we were both members of the Royal Society Education Committe under the chairmanship of Clifford Butler and later Harry Pitt. Of particular interest during this period was a Royal Society discussion paper 'Science and the Organization of Schools in England' prepared on behalf of the Education Committee in 1979 by Nevill Mott after an extensive series of visits by members of a working party to schools, colleges, university departments of education, etc., throughout the United Kingdom. Clifford Butler refers (p. 191) to this activity and the main point that I would like to make is that Nevill Mott and the working party then considered that

It [is] of the highest importance that separate courses in physics, chemistry and biology should be retained in as many schools as possible, taught by specialist teachers and open to pupils with the ambition to become scientists, engineers and technicians.

and that

Also we do not believe in forcing a talented child to study a subject that he (no she!) feels is of no use to him.

These views (with which I then associated myself) were strongly endorsed by Nevill Mott when he spoke at a Royal Society meeting in November 1979 to discuss this paper and one entitled 'Alternatives for Science Education', which contained contrary views from the Association for Science Education. Nevill Mott's position was somewhat at variance with the integrated features of science degree courses that he had envisaged for the University of Sussex. Nor was it later sustained by the Royal Society in a report produced in 1982 by a study group, of which he and I were members:

We are convinced by the force of the arguments that the science curriculum up to sixteen for *all* pupils should be balanced in that it covers work in physics, chemistry and biology. We are particularly anxious that a pupil should not be forced to choose among the branches of science thereby becoming excluded from further study in one or more subjects.

These views have been sustained by the Royal Society during the last fifteen years and are indeed essentially those embodied in the present National Curriculum. How such science courses should be delivered, as integrated or separate sciences, and whether they should occupy, 10%, 20% or even up to 30% of curriculum time for different pupils is still a matter of discussion and argument. But the Royal Society has had considerable influence in developing the current views and there is no doubt that Nevill Mott's input into its educational work over the years has been of great and lasting importance.

PROFESSOR R. J. BLIN-STOYLE, FRS, is Emeritus Professor of Physics at the University of Sussex. His appointments include Lecturer in Mathematical Physics at the University of Birmingham, Senior Research Officer in Theoretical Physics at the University of Oxford, and Fellow and Lecturer in Physics (now Honorary Fellow) at Wadham College, Oxford. The Founding Science Dean at Sussex University, from 1962 to 1990 he was its Professor of Theoretical Physics. Other positions he has held are Chairman of the School Curriculum Development Committee (1983–88), President of the Institute of Physics (1990–92) and President of the Association for Science Education (1993–94).

Student Rebellion and how NEC got a Grant

David Baron

Sir Nevill was Chairman of the Trustees of the National Extension College (NEC) in Cambridge for much of my time as Executive Director (1967–1971) and I have a vivid recollection of his impressive presence, melting into great charm and sympathy.

This was the period of student rebellion, from which the Cavendish Professor was not immune. For instance, he once remarked to me that it was essential for him to listen, in person, regularly and often, to his undergraduates' grievances and complaints and to answer them effectively; otherwise 'they might well have smashed up the laboratory.' Similarly, at Birmingham University, the Registrar arrived one morning to find her office and filing cabinets broken into and students' confidential personal files strewn about the floor. 'You must surely have had a very uncomfortable night,' she said to the perpetrators, 'I shall make you some tea.' The perfect emollient!

During this time there occurred the Garden House Affair, when Cambridge students invaded a riverside hotel and interrupted a lunch organized by the Spanish Embassy (during General Franco's regime) to encourage tourists to visit Spain. The students broke some windows and did other damage in the hotel, disrupting the lunch. Sir Nevill told me succinctly how it was a very good thing they went for the undistinguished Garden House; otherwise they might well have attacked the beautiful Senate House!

The National Extension College, a correspondence college distance-teaching in collaboration with the Open University and the BBC, was in those days chronically short of funds and the Trustees decided to apply for a government grant. I was present 'below the salt' during two meetings, at the first of which the case was presented by the Trustees, under Sir Nevill's chairmanship, to the Secretary of State for Education in the Labour government. Mr Short promised to consider the application, but nothing happened; the case presented was on the lines that the NEC was offering an important educational service, designed to encourage people who would not normally have aspired to graduate, by providing preparatory courses leading to an Open University degree. A general election ensued and the Trustees then applied a second time, appearing before the newly appointed Secretary of State, Mrs Thatcher. To her, the case for grant subsidy was presented quite differently: a valuable educational institution and a fine example of private enterprise, self-financing and independent, suffering temporarily from lack of resources owing to its ambitious initiative in leading students up to degree level at the Open University.

The Secretary of State responded at once, to the consternation of her Private Secretary, as I thought, by saying at the meeting that since the Treasury was already financing the Open University to the tune of some millions of pounds, surely a few thousand could be found for the NEC. Quite soon afterwards we were telephoned to say the college had been granted £30 000.

MR D. W. BARON, OBE, was born in 1915 and educated at Winchester and Oxford. He served in the Colonial Service (now HM Overseas Civil Service) from 1937 to 1966, first in Ceylon then with the East Africa High Commission. His final postings, in Hong Kong, were as Deputy Colonial Secretary and Director of Social Welfare. He was the Executive Director of the National Extension College from 1967 to 1971.

Open Learning

Michael Young

I forget how I first met Nevill; I think it was through Brian Jackson, who joined me in founding the National Extension College in Cambridge. I could hardly believe our luck. The Cambridge Establishment had been adamantly hostile to the kind of proposal we were making for an Open University. Our magazine, *Where?*, which was edited by my wife, Sasha, in 1963 contained the first proposal for an Open University and used that name instead of the University of the Air. Showing how green we were, we started off by suggesting the new university might be a second University of Cambridge, using the colleges and the Cavendish in the vacation but with correspondence courses and cassettes in term time. To begin with, our supporters could be numbered on the fingers of one hand.

One of the main objections was that the vacations were the time when the dons did their 'real' work, their research and writing. They could not simultaneously teach a whole new second lot of students. So we trimmed our original proposal: outside staff would be brought in to teach during the vacation. We got the agreement of the then Battersea Poly to do this. The idea of physicists from the poly supervising practical work in the Cavendish was more greatly detested than the rest of what we were proposing. It was anathema to almost everyone, but not to Nevill; he was unusual in so many ways.

From then on he was a supporter of the National Extension College (set up in 1963) in its role as a pilot project for an open university, and of the Open University (OU) when it happened. Even though he was the Cavendish Professor and Master of a college, he was unlike most of the elite of the university in being almost as much concerned about people who had been deprived of higher education as he was about the more fortunate who had made it to Cambridge.

A little later on, when the Open University was about to become a fact, and the amazing Walter Perry had been appointed its first Vice-Chancellor, Nevill invited Perry to dinner at Caius College to meet the NEC trustees. As I remember, Perry was then thinking the OU should have its headquarters in London. With Nevill as the genial host he always was, I think we persuaded Perry that it would be much more appropriate for a new university, and a new kind of university without internal students, to be in a new town like Milton Keynes, which had the additional advantage of being at the geographical centre of the country. Perry took up the idea immediately, and within a few days negotiations had started with Jock Campbell, the Chairman of the New Town Corporation. Campbell could hardly believe his luck.

When I retired from being Chairman of the NEC, Nevill took over for the period 1971 to 1976. What do I remember of him? How much I liked him, to start with. I can't think of many people in my life for whom the word 'benevolent' was so truly apt. He shone with benevolence and a similar quality, goodness. More than merely seeming to like the odd lot of people we had got together, I am sure he truly did like them and was amused by them.

When he disapproved of something that was being proposed, he certainly made his view plain. He did not suffer foolishness gladly. But even when he was exasperated, he was so nice about it that no one could take offence. His childlike quality – which I believe outstanding people usually have – was also endearing, as was his willingness to support a new Cambridge college that was definitely outside the fold.

Nevill gave a sense of confidence to the small staff of the NEC that they otherwise could have lacked, and his close interest in the detail of a distance-teaching college was a source of great encouragement to everyone. He would have been pleased that the National Extension College is now so flourishing under the direction of Ros Morpeth and the chairmanship of Geoffrey Hubbard, and that it has found a very sizeable place for itself alongside the Open University.

LORD YOUNG OF DARTINGTON is a sociologist and Director of the Institute of Community Studies. He is an Honorary Fellow of the British Academy and Churchill College, Cambridge. He was the originator of the Open University and founder of the Consumers Association.

Teaching Science

Clifford Butler

In his autobiography Nevill Mott tells us that, when he became Cavendish Professor of Physics at Cambridge in 1954, he was persuaded by Sir Philip Morris, the Vice-Chancellor of Bristol, to devote time to national educational issues. He joined Geoffrey Crowther's Committee on Education from 16 to 18 in 1956, and in 1959 the Ministry of Education's Committee on the Training of Teachers, chaired by John Fulton. Nevill was invited to take the chair of a subcommittee principally concerned with the shortage of mathematics teachers. Thus he began a series of commitments to educational issues which lasted into the 1980s.

I first worked with Nevill in 1961, when we both became members of an advisory committee concerned with the Nuffield Foundation's programme for the development of a new A level physics course. Nevill chaired this small group of academics and teachers. We soon realised how difficult it was to construct a curriculum to meet the needs of pupils who would subsequently enter a wide variety of degree courses. Nevill and I wanted the new course to concentrate on classical physics, but the teachers argued that it must include quantum physics, particularly for the majority of the pupils who would not continue with physics in depth. Debate between the school and university worlds ultimately made a valuable contribution to the great success of the Nuffield physics course.

From 1965 until 1971 Nevill was chairman of the Physics Education Committee set up jointly by the Royal Society and the Institute of Physics. During this period the main concern of the committee lay in teacher training. Late in 1969 the Royal Society established a Standing Education Committee. I chaired this committee until the end of 1979 and Nevill was a member until 1982. The range of our work extended from primary schools to postgraduate courses as well as coordinating the work of four ongoing committees for specific subjects – biology, chemistry, physics and mathematics – each of which was run jointly by the society and the professional body for each subject.

It was not easy to find Fellows of the Royal Society (FRSs) who were knowledgeable about a wide range of educational issues and willing to devote a considerable amount of time to promote any change. Nevill Mott, however, was an exception. He was at his best working on his own, or perhaps with a few colleagues, but he always enjoyed the help of the society's education officer, Donald Harlow, a former successful schoolteacher. Over the years, Nevill chaired several ad hoc groups studying specific problem areas. Reports were usually published after approval by the Standing Education Committee and the Royal Society Council. I served on some of these groups and can testify that Nevill was a conscientious chairman who spent a lot of time on the committee's business between meetings. He particularly liked to visit schools and colleges for informal discussions with teachers on their own ground. Don Harlow used to arrange these visits and occasionally I would join them. This ensured that we had some first-hand experience of the problems under consideration.

The Standing Education Committee's reports were circulated widely among appropriate professional bodies and sometimes national conferences were organised to promote discussion. Nevill several times acted as chairman, with great charm and persuasive power. Occasionally the Royal Society prepared papers for wide circulation in schools. One of its bestsellers was a booklet on metric units in primary schools; eighty thousand copies were sold between 1969 and 1973.

I now give some example of the studies undertaken by Nevill on behalf of the Standing Education Committee.

In 1968 the Council of Engineering Institutions and the Royal Society instituted a study of the supply of schoolteachers in science and mathematics. The conclusions, published in 1969, showed there were significant shortages of mathematics and physics teachers. It was decided that the Royal Society should collect statistics on these shortages on an annual basis and discuss them with the government. Nevill often accompanied me on these visits to the Department of Education.

The new Standing Education Committee set up a study group consisting of eight FRSs who were invited to make recommendations concerning the training of science and mathematics teachers. Nevill was chairman of this important group. Five working parties, all involving practising teachers and trainers as well as FRSs, were set up to provide detailed information for the final report. A large consultative group, with a widely based representation, was also convened to advise as the work progressed. I remember that Nevill devoted a substantial amount of time to this study. He visited schools, teacher training colleges and university departments of education to familiarise himself with current practice and for informal discussions on how to improve both recruitment and training in the future. The final report, *The Training of Teachers of Science and Mathematics*, was published in 1972. It contained fifteen specific conclusions and recommendations. Well received, it undoubtedly made a worthwhile contribution to the ongoing debate during the 1970s.

During 1972–74 Nevill Mott was concerned with environmental science courses for students intending to teach science to 11–13 year olds. Again he visited colleges and schools before preparing a report for the Royal Society. In 1976 he was concerned with primary science and in 1977 he became deeply concerned about the needs of talented children.

A small study group was set up to support and advise on Nevill's personal researches into the controversial issue of whether special support was needed for talented children in the state sector of education. As usual, Nevill made numerous visits to institutions and consulted widely before his group issued a discussion paper in 1979, 'Science and the Organisation of Schools in England: Implications of the Needs of Talented Children'. In his autobiography Nevill explains that some members of his committee felt that talented would-be scientists might not receive the attention and stimulus they needed in every comprehensive school. Mixed ability teaching was coming into favour towards the end of the 1970s and priority appeared to be given to the needs of the average child. Furthermore, there was controversy between those who believed a single integrated science course was preferable to three separate courses in biology, chemistry and physics. Nevill also points out that the Association for Science Education, in a paper called 'Alternatives for Science Education', seemed to favour a pattern of education which appeared to his group to be more about science as a social activity than a job of work.

Nevill's paper considered the suitability of mixed-ability teaching methods in science, particularly for talented pupils. Streaming and setting arrangements were also discussed. After careful consideration the group favoured setting in science and mathematics from at least age 13 onwards in non-selective schools. He also hoped that the separate sciences would continue to be available in most secondary schools. Concern was expressed that falling numbers of pupils might jeopardise these recommendations.

In November 1979 the Royal Society held a one-day discussion meeting on science education in secondary schools to decide priorities for the 1980s. In one of the main talks, Nevill described the Royal Society's view of the needs of talented pupils with a variety of ambitions and abilities. Although some of those present felt that his position was too elitist, it seems to me that the discussion was timely and may well have influenced future events.

As a further contribution to the national discussion of the requirements of high-ability children, and with the help of the Leonardo Trust, Nevill organised a weekend conference on gifted children and their contemporaries. It took place in Cambridge during September 1981 and I was one of the many educationalists who attended. Being well aware of the view that his report was elitist, Nevill made a well-balanced and conciliatory speech at the beginning, the text of which is reproduced in his autobiography.

I now look back to the 1970s and to all the educational activities I shared with Nevill Mott, a few of which I have mentioned here. Education at all stages in the development of individuals is a large-scale and complex process. It is very rare for one person or even a committee to have a revolutionary effect on developments; change more likely comes about by slow evolution. Certainly our reports did not lead to dramatic changes, but it seems to me now that we must have influenced the direction of change. Several of the issues to which Nevill Mott devoted valuable time have continued to attract attention as circumstances change. In particular, I have noted recent concern by some politicians about the ongoing needs of talented children. Without doubt, Britain still needs to maximise the output of talented people from its education system.

I certainly believe that Nevill's contributions in the field of education were very worthwhile. He presented his carefully thought-out opinions effectively, but he was always ready to acknowledge legitimate diversity of opinion. Nevertheless, I think he usually reached the right conclusions in his carefully prepared lectures and reports.

SIR CLIFFORD BUTLER was educated at Reading School and Reading University. He was a staff member of the Physics Department at Manchester University (1945–53) and Imperial College (1953–70). Thereafter, he was Director of the Nuffield Foundation (1970–75) then Vice-Chancellor of Loughborough University (1975–85). He represented the Vice-Chancellors Committee on the Schools Council (1965–84).

Taylor & Francis

My Memories

Bryan Coles

The range of Nevill Mott's interests was so great that there must be a cloud of witnesses to the stimulus he gave to younger people.

I was never his student or coworker but no one had more influence on my work and my career. It began at a prize-giving when I was a metallurgy student in Cardiff; I received a copy of Mott & Jones, which I still treasure. Much of it I did not understand then, but my supervisor, Hume-Rothery (H-R), in what was then the Department of Inorganic Chemistry at Oxford, encouraged my awakened interest in the magnetic and electrical properties of alloys. As a result, when I had moved to Imperial College to lecture in metal physics and Nevill had asked H-R to review the transition metals and their alloys for his new journal *Advances in Physics*, Nevill agreed to let H-R have me as a joint author on their physical properties. That was of great help to me as I then spent the next two years in the United States, working for the first time with other experimental solid-state physicists, and discovered that one did not have to be reared in the Cavendish or the Mond to use liquid helium. Much of the work I did there was inspired by clues drawn from Mott & Jones, and on my return, Nevill asked me to write what became my second *Advances* article. It was on spin-disorder scattering, a topic which Nevill realized had to be more appropriate for explaining the resistivity of gadolinium than the model he had put forward for nickel twenty years earlier.

By then P. M. S. Blackett was head of the Physics Department at Imperial, but I think he only realized he had inherited a metal physicist when Nevill, visiting his old friend, indicated that he was interested in my work. Later, in 1963, when Nevill needed an editor for *Advances* (Brian Flowers having decided he could not convert it to a journal of nuclear physics, and many of the articles being still in solid-state physics), he invited me to do the job. That job I did for thirteen years, constantly helped by Nevill and by the hopes of authors that he would become aware of their work.

Nevill was by then the guiding spirit of publishers Taylor & Francis, setting the company on a road of growth that has continued to the present day. He recognized that a scientific publishing company needed the real involvement of members of the academic community it sought to serve, and I was honoured to be invited by him to become a non-executive director and later his vice-chairman. I do not think I would have dared to accept the succession, when he stepped down as chairman, had he not accepted the role of honorary president, making himself always available with wise advice.

At Imperial College, too, he was a great source of strength as a senior Research Fellow after his retirement from the Cavendish, helping individual students and assisting the Solid-State Physics Group to acquire visibility in the shadow of the long-established major groups in other fields.

Nevill records in his autobiography that the greatest pleasure of a life in research is putting others on the way to success. He certainly did that for me and many besides. Furthermore, he never gave the impression of doing it *de haut en bas*; I will always remember that when a graduate student in a northern university was doing work that interested him, Nevill did not summon him to his presence but wrote (in that tiny handwriting we all remember) to ask, 'May I come and visit you?'

PROFESSOR BRYAN COLES, FRS, died on 24 February 1997 aged 70. He was Professor of Solid-State Physics at Imperial College for twenty-five years and Pro-Rector from 1986 until his retirement in 1991. His research interests were in magnetism and related phenomena in metals. He edited Advances in Physics *for thirteen years and served as Chairman of the Board of Taylor & Francis from 1981 to 1995.*

Editor, Chairman and President: three roles in one Company

Elnora Ferguson

The publishing firm of Taylor & Francis began as a small family business to print *Proceedings of the London Philosophical Society*. From these small beginnings in 1798, Taylor & Francis has grown into an international publisher of stature and the *Philosophical Magazine* into a leading world physics journal. This transformation of the firm began in the years when all aspects of the country's life and work had to be rebuilt after the Second World War. Sir Nevill Mott was closely involved in many aspects of this development.

Professor Mott was at the University of Bristol when he was approached to take over the editorship of the *Philosophical Magazine* from Professor Allan Ferguson, whose health was deteriorating. Mott became editor in 1948, with Ferguson remaining as reviews editor until 1951, when Mott took over control of the whole journal. His editorship continued until 1970. During this time he was responsible for many developments: the establishment of an editorial board, formalising the previous arrangement of an editor with one or two assistants; and discussions with the Physical Society to try to ensure that UK physics journals complemented one another, instead of competing in their research coverage. Partly as a result of these discussions, the *Philosophical Magazine* moved into Professor Mott's own research area of solid-state physics. The submission early in the 1950s of a lengthy article led to the conclusion by the editorial board that there was the need for a review supplement that could publish material of length. This proposal to Taylor & Francis led to the new journal *Advances in Physics* in 1952. A second review journal, *Contemporary Physics*, established in 1959, was designed to reach schoolteachers and undergraduates.

Some members of the postwar board of directors died during the 1950s, members who had initiated the transformation from family firm to its current status, and it was hoped that Professor Mott would become a director, but he had been appointed Cavendish Professor in 1954 and did not have the time to be even more closely involved with the firm. Although not a director, his editorial work helped him to initiate structural changes in management techniques to improve effectiveness as the company grew.

In 1967, five years after his knighthood, Sir Nevill agreed to become a director, and in 1970 he became chairman. During these years of his membership of the board, the structures were strengthened which have enabled the firm's growth and development over the past thirty years, so that Taylor &

Francis became recognised as a publisher, not as a printer who also published. In the reorganisation of the 1940s, the board of directors had been established with equal numbers of scientists and non-scientists. Sir Nevill ensured this balance continued.

His retirement in 1975 did not indicate his withdrawal from the company and its work. He was made Honorary President, a specially created post, and remained a regular attender at board meetings until well into the 1980s. He followed all the company's activities closely and was a splendid ambassador in all his activities. Letters, written in his own hand and frequently by return of post, continued to arrive until a few days before his death.

Taylor & Francis owes much of its current success to the contribution made by Sir Nevill Mott during fifty years of close association. Within the company, he will long be remembered with affection for the personal interest he took in everyone involved in the work, from his fellow directors to the newest member of the office.

MRS E. FERGUSON is currently Chairman of the Board of Taylor & Francis. She is a social scientist and statistician and daughter-in-law of Professor Allan Ferguson, a former director of the company and editor of the Philosophical Magazine.

Late Cambridge

Reminiscences

Sam Edwards

In 1958 I was working in Rudolf Peierls' department in Birmingham; I had become tired of nuclear physics and asked Peierls for a problem in solid-state physics. He suggested I should calculate the resistance of a metal at absolute zero, since he did not believe the work of Pomeranchuk who said it would be zero. Peierls and Derek Greenwood had derived what are now known as Kubo relations and were trying to evaluate them. I recognised that these were Green functions and quantum field theory methods, already in existence, could and did solve the problem. The solution seemed to go far beyond this particular problem, so I decided to show the method to Nevill Mott, who was of course the leading solid-state theorist in the country. I set off to Cambridge where I met him for the first time. I explained the work to him and asked him where he felt new theoretical methods were required, suggesting alloys as a possibility.

He was polite, but totally sceptical, and said that in his view the next advance would be in liquid metal theory. This, as ever, was most percipient, and the Cavendish theory group of the time produced a series of splendid papers, resolving the striking but hitherto unexplained changes in the properties of metals on liquifaction. I did myself follow Nevill's suggestion and I solved several problems, but the cream was taken by his immediate colleagues and is admirably developed in Tom Faber's *An Introduction to the Theory of Liquid Metals*. There was, however, a sequel. The Green function formalism is very good to write down solutions in abstract exact form, which gives unassailable answers when used in comparatively simple situations. So I was very pleased when, a year or two later, Nevill confessed himself puzzled by the way the density of electronic states came into some final formulae and I was able to help by showing that the cancellations he somehow knew had to happen, did actually happen.

He went on to think about Anderson localisation and, I think much influenced by Borland's numerical work, argued for the universality of localised wavefunctions by arguments which struck me as simplistic. (The trouble I always had with Nevill was that he presented things in such simple terms that I always felt unconvinced; but he was always right.) It took me years to be convinced on this matter, but convinced I was in the end.

Nevill became Master of Gonville and Caius College, and I am a Caian, so during my time at Manchester I occasionally saw him in that role. A notable visit had to do with the Pugwash movement when a lively discussion in the lodge at Caius with Brian Flowers and John Maddox remains in my memory. Nevill was not much interested in politics as such, but was by nature conservative and uninfected by the unreasonably gentle view many academics had of the Soviet Union *et al.* So it was very important that people of his standing were in Pugwash, to convince government that it was not some kind of front for fellow travelling views. He, like Rudolf Peierls, was completely realistic in this, appreciating how the magnitude of the incipient disaster of nuclear conflict fully justified the attempts of Pugwash scientists to keep a bridge between the nuclear powers.

He later became very tired of the college and resigned the Mastership. It was said that this move was precipitated by his failure to get the college to run its finances with professional advice, perhaps from a merchant bank, electing instead the University Lecturer in Tibetan to be its bursar. Be that as it may, when I returned to Cambridge in the seventies, after I had established myself in the college, I was very pleased to welcome him back to the beautiful room which Caius had assigned to its physics professors and, in this way, I met many of the interesting people he brought into college. He was a brilliant linguist and I confess to being terrified when meeting the many French scientists who visited; Nevill expected me to continue the conversation in French. He loved to offer his visitors malt whisky, perhaps to see their reaction, but more likely to have an excuse to enjoy it himself.

We did not talk much physics in this later period, for my interests had turned to chemical physics and his to superconductivity, but I was able to help in persuading the college to underwrite the position in Cambridge of Sasha Alexandrov, which proved a great help to Nevill in his last decade. The IRC in Superconductivity was awarded to Cambridge and (for reasons I have never quite understood) the Research Council gave Cambridge a new building for it, to which Nevill then moved. Thus the head of the Cavendish had a splendid stick with which to beat researchers who objected to their space allocation, for he could then say that Mott had been expelled from the Mott Building, so what did they have to complain about?

Nevill continued to visit the Cavendish, the IRC and Caius right up to a week before his death, and he always had interesting comments on any bit of physics he was asked about. A truly awesome person.

PROFESSOR SIR SAM EDWARDS, FRS, obtained his first degree at Caius College, Cambridge, and his PhD at Harvard; he then worked with Rudolf Peierls in Birmingham. He moved to Manchester in 1958 then to Cambridge in 1972. He became Cavendish Professor on the retirement of Sir Brian Pippard, who had succeeded Sir Nevill Mott. He collaborated with Nevill on the theory of liquid metals and semiconductors; his current interests are in polymers.

A Happy Warrior

Phil Anderson

As was the case with many of the great figures of twentieth-century physics, Nevill Mott changed from legend to reality for me at the Kyoto International Congress of Theoretical Physics in August 1953, the first considerable conference held in Japan since the war.

At this time I had been a solid-state physicist only four years, and for me Mott had been a totally remote figure, one of the earliest originators of the field, from whom even my mentors had learned the canon. I was present more or less by accident, whereas Mott was a very senior panjandrum indeed, being the chair of the International Union of Pure and Applied Physics (IUPAP), the conference sponsor. But my four-year-old daughter, Susan, was photogenic and occasionally shared the photogravure pages with Mott and Hideki Yukawa.

Our actual acquaintance began on a long train journey from Sendai to Hokkaido, on which Mott was extraordinarily patient with Susan. We even have a photograph of her on his lap during the train ride.

I don't remember what little science we talked then. Much more scientifically fruitful were a couple of his visits to Bell Labs in 1956–57, during the period when he was working on his second paper on the Mott insulator, and we at Bell were fascinated by the new high-quality phosphorus-doped silicon we were studying at low temperatures, with magnetic resonance, thermal, transport and optical probes. Though there was some precedent in work of Vonsovsky and remarks of Peierls, Mott gave us all our present clear picture of the interaction-dominated insulating state, and in discussions among Herring, Lax, Lewis, Wolff, and myself, and later Kohn and Luttinger and other summer visitors, we came to attribute the insulating nature of the lightly doped impurity band to Mott's phenomenon. (Mel Lax in particular made this point.) This attribution was why I hedged my bets by emphasizing localized spins in my localization paper, which was in preparation at the time. The electrons, I supposed, could always be localized by Mott, but spins could not be localized by Coulomb forces.

By the time I wrote my second superexchange paper (fall 1958), I had completely internalized Mott's arguments, and the formalism of this paper was based on what came to be known as the Mott–Hubbard Hamiltonian. I considered it the direct descendant of Mott's 1949 and 1956 papers.

When I came to spend a sabbatical in the Cavendish in 1961–62, I was housed in Hartree's old office near Mott's on the second floor in the Austin

Nevill with Phil Anderson.

Wing, and he and his student Borland would every once in a while pin me down for a chat about localization. It was in the ebullience of a dinner celebrating his knighthood, that year, when he first mentioned, very publicly, the possibility that I might eventually come to Cambridge. (This changed the attitudes of some staff members considerably.) Joyce and I also remember a delightful formal dinner in the Master's Lodge at Caius, where our Midwestern table manners and customs were severely tested.

It was characteristic of Nevill that he had an extraordinarily broad and eclectic spectrum of contacts at Bell, as elsewhere, and he would drop in from time to time as the guest of various scientists barely known to us in the theory department. It was on one of these visits, five years later, that he suggested my coming to one of his newly created professorships. In due course my status as permanent visiting professor was approved and from 1967 to 1975 I spent eight productive years in what those who didn't have to make the housing arrangements invariably called the best of both worlds.

This was the period when Mott was putting together, with Ted Davis, Hellmut Fritzsche, and other members of what has been called his worldwide laboratory, the evidence which eventually convinced people that semiconductors did indeed show both an Anderson and a Mott transition. Deeply involved in various other problems, I could only look on, in wonder, at Nevill's unbelievable level of activity during those years. Though he had given up the Mastership of Caius, he was still full-time as head of the Cavendish, which contained not only Lord's Bridge but also the powerful MRC unit, soon to depart for south Cambridge. For several years I sat on the Faculty Board of

Physics and Chemistry and observed with delight the exchanges which inspired the famous couplet about Todd and Mott.* He was also active in Pugwash and was pushing schemes on elementary education. In spite of all this, concepts (variable-range hopping, the mobility edge, the minimum metallic conductivity), papers and books poured out, and he seemed as well to give a featured talk at every meeting anywhere around the world which would touch on his beloved subjects of amorphous and disordered systems, and metal–insulator transitions.

Every term he and Ruth would invite all of scientific Cambridge to a sherry party. Joyce learned early to beware the typical Mott introduction, 'Beryl, this is Joyce Anderson,' leaving Joyce to find out for herself that this was Dame Beryl, Mistress of Newnham, and one of the powerful figures of Cambridge – presumably Dame Beryl was equally bewildered. I also remember a clipped remark on the way to a lunch at Caius. I was telling Nevill that my college was experiencing a Mastership election worthy of a C. P. Snow novel; he quickly rejoined, 'He didn't know the 'arf of it!' His only allusion ever, perhaps, to his struggles at Caius.

After retirement as Cavendish head in 1973, Mott became even more formidable in terms of his energy, his wide acquaintance throughout the world, his readiness to publish, even in book form, almost before the opposing point of view was available as a paper, and his delight in attending every conference at which the much younger crowd attracted to the fields he created would convene. The Mott concepts remained central to the active fields of electron-hole droplets, of the Mott transition in V_2O_3 and other transition metal oxides, to the early work on inversion layers which led to the quantum Hall effect, and to problems of electron transport in SiO_2 and other oxides. He barely paused to pick up his Nobel prize in 1977 (and remark happily that he was glad he was not the first Cavendish Professor in the twentieth century not to receive one) before immersing himself in the controversy over whether there could be a scaling of conductivity. And throughout this period he was nursing along his dream of a technology of amorphous semiconductors.

It was characteristic of Nevill that he was completely devoid of intellectual 'side' in his choice of associates. He had been there at the birth of the quantum theory and he appears in the Solvay Conference photographs of the twenties and thirties, behind Marie Curie, Einstein, Rutherford, Pauli and the like; and within him one sensed a conscious decision to go his own way and to do science in the practical service of mankind, and to accept support for that science (but never for personal gain) from any available source.

It is more than any single person can do to appreciate his wide variety of activity in a small space. Perhaps I can only call attention to specific contributions to fundamental science which may be lost in the welter of practical applications. The Mott transition is still a very deep and active problem, not really

* Lord Todd/Thinks he's God,/Nevill Mott/knows he's not/and Mott made Master faster.

wholly solved or understood, with ramifications in unexpected fields like quantum chromodynamics. Nor, for that matter, is the Anderson transition: the recent discovery of a critical point in inversion layers is a really deep puzzle. Mott and Pepper, incidentally, tied with the Bell group in their realization that these were the ideal two-dimensional systems; this underlies the whole further development of the quantum Hall effect, which in turn relies heavily on Mott's concepts of localization and Coulomb interactions. The crucial role of the minimum metallic conductance, e^2/h, was emphasized by Mott years ahead of anyone else. Some Germans have proposed this unit should be named after von Klitzing, but in my mind it came into physics through Mott.

My best memories of Nevill, and of course the most immediate, date to the post-retirement period. He would arrive at a workshop at some Alpine resort – Aussois, Morgins, Les Arcs – and be surrounded by a group of physicists averaging half his age, eager and ready to do battle over the latest topic of controversy. Nevill took on all comers in discussions, which never – well, almost never – had the slightest trace of acerbity, and were always open and constructive. (Not that it was ever easy to persuade Mott.) I think of him as a happy warrior, glorying above all in the creative rough and tumble of scientific life.

PROFESSOR P. W. ANDERSON obtained his BS in 1943 and his PhD in 1949, both from Harvard. He worked at the Naval Research and the Bell Telephone Laboratories before being appointed to the Joseph Henry Chair of Physics at Princeton University in 1975. He was Visiting Professor of Theoretical Physics in Cambridge from 1967 to 1975 and the George Eastman Professor of Physics at Balliol College, Oxford from 1993 to 1994. He has been the recipient of numerous honours and prizes, including the Nobel prize for physics, which he shared with J. H. Van Vleck and N. F. Mott in 1977. His research interests cover a broad range of topics in theoretical physics.

Some Personal Recollections

John Enderby

I think it was in 1958 that I first met Nevill Mott. I was working on my PhD under the supervision of Norman Cusack on the experimental determination of the Hall coefficient of liquid metals. This proved to be a difficult measurement to make, and even today reliable results are few and far between, particularly at high temperatures. Nevill had been to Birkbeck College earlier that year to talk to Norman, and with the growing interest at the Cavendish in disordered matter, he had decided to organise a small invitation-only conference. As a very junior and insignificant member of the scientific community, I was amazed that at the end of one session he sought me out and asked my (yes my) opinion as to what was the likely outcome of the Hall effect experiment. I stuttered some reply – I can't remember what it was – but after the meeting he wrote me a kind note encouraging me to continue the measurements and saying he would read any paper I wrote with considerable interest.

People who knew Nevill will of course recognise this wonderful gift; although a professor of international fame and reputation, he was able to interact with scientists at all levels, irrespective of their background, affiliation or reputation. He was at ease with first-year graduate students and Nobel prizewinners, and could talk to both on equal terms. He never patronised the former or flattered the latter!

Nevill organised a second meeting a year later. This was memorable for two reasons. First, John Ziman outlined his theory of liquid metals – a theory which revolutionised the subject and is now taught at undergraduate level. Second, a visiting American professor, Dr Phil Anderson, gave an account of what happens to the diffusion of spins on a random lattice. I would love to say that I recognised that here was an approach of seminal importance. Alas, I did not and nor, I suspect, did most of the audience. Whether Nevill followed Anderson's methods in detail, I know not; what I do know is that his intuitive grasp of quantum physics enabled him to realise that this was indeed a major breakthrough. I think most people would now agree that Nevill, more than anyone, showed the value of Anderson's work to the wider community; so much so that the *Physical Review* paper which describes the study is often dubbed 'the most quoted but least read.' The Mott–Anderson transition has, of course, dominated the thinking of condensed matter physicists interested in disorder for over three decades.

I remember, too, a third meeting at Cambridge held in 1962. This occurred just after his knighthood was announced and the news was received by all attending with considerable pleasure. The then Minister for Science, Lord Hailsham, wrote to Nevill congratulating him on the award; the letter said it was in recognition of his services to *experimental* physics. I think it was Brian Pippard who commented that this was rather more appropriate than the minister realised! Nevill always took an interest in the latest results of an experiment and his careful, critical but sympathetic analysis of one's interpretation invariably stimulated further activity. His enthusiasm for the subject never waned and was truly infectious.

Years later, when I was at Bristol, the news of his Nobel prize came through and I, like hundreds of others, wrote to offer my congratulations. When I was appointed to the H. O. Wills Chair at Bristol (a chair that he himself had occupied for many years), Nevill was one of the first to send me his best wishes. I was thrilled and honoured that in his book *A Life in Science* he chose my letter to print in full as he described his winning of the Nobel prize.

He was a great man and to me a personal friend of incomparable warmth who, at every stage in my career, offered help and guidance. I count myself as privileged to have known him; we all miss him.

PROFESSOR J. E. ENDERBY, CBE, FRS, is the former H. O. Wills Professor of Physics at Bristol, a chair which Nevill held for many years before moving to Cambridge. His interests are in the properties of liquid metals, semiconductors, molten salts and aqueous solution and he was awarded the 1995 Guthrie medal and prize of the Institute of Physics for his work in this area.

Working with Nevill Mott

Michael Pepper

In 1971 Sir Nevill Mott retired from heading the Cavendish and became a consultant to Plessey, the electronics company where I was working. With hindsight one can realise that it was a most propitious time as the semiconductor technology was developing rapidly. Silicon (Si) was on the verge of moving into the era of very large scale integration (VLSI) and the properties of the silicon–silicon dioxide interface and silicon dioxide (SiO_2) itself were of the greatest importance for the successful development of integrated circuits with a very high transistor packing density. In addition the technology of the growth of III–V semiconductors was also developing rapidly. At that time, Nevill was piecing together the physics of amorphous materials, looking for examples of localisation in a wide range of disordered systems. He became a regular visitor to the laboratory at Caswell, near Towcester, Northants, where his visits were coordinated by Chief Scientist Bryan Wilson and Director Derek Roberts, who subsequently became Deputy Managing Director of GEC and is now the Provost of University College, London.

We discussed the problems of charge trapping and interface states in the silicon–silicon dioxide system, which determine device performance and reliability. As he considered the work and looked at the experimental literature, in his characteristic way, he was able to extract the fundamental aspects of the problem from what seemed a very confused and complicated situation. It became clear that both his present and past work on disorder in general, and the properties of ionic crystals as outlined in the classic text, Mott and Gurney, were highly relevant. He recognised that the formation of the U centre in the alkali halides, which in the late 1930s he had shown to be a negative hydrogen ion substituting for a halide ion vacancy, was the same process as the removal of surface states at the Si–SiO_2 interface by annealing in hydrogen. Ejection of the trapped electron by illumination, or application of a high electric field, possibly accompanied by hole injection, allows the resulting hydrogen atom to diffuse away. The net effect is the return of the surface state and device instability.

The question of electron transport through silicon dioxide is of importance for the operation of non-volatile memories in which carriers are injected into the oxide and then trapped. Photo-injection experiments had shown that the electrons possessed a surprisingly high mobility with little trapping. This was most surprising to us, after all the oxide was an almost prototypical amorphous material. Nevill realised that the physics here is the same as the physics behind a high electron mobility in the liquid rare gases. Thus if the bottom of the conduction band is s-like, then an electron wave function experiences an averaged potential with little difference between the crystalline and amorphous states, and a consequent reasonable value of electron mobility. Another area which interested him was the interpretation of some data obtained by a group at AT&T, who were investigating Al_2O_3 as the gate dielectric of MOS transistors. They deposited a layer of Al_2O_3 on the thermally grown SiO_2 and found there was a change in the threshold voltage of the device; the change was independent of the thickness of the SiO_2, so it was difficult to reconcile with charge at the interface of the two dielectrics. This he readily explained as being due to an effect rather like a work function arising from peaks in the interface state density at the boundary of the two dielectrics. The Fermi energy was continuous at the temperature of formation, so the peaks were aligned, producing a constant voltage across the SiO_2 regardless of the thickness of the layer. This was a revelation to those of us who were at that time working at the frontier of technology, and he derived enormous satisfaction from piecing together these and other findings then fitting them into the framework he had developed. Indeed, in an article which he wrote a few years later, he stated that some of the most fascinating problems in solid-state physics were those thrown up by technology.

It was in the area of electron transport that his interest had the most influence. One day early in 1972 I showed him one of the most notable papers on conduction in the two-dimensional electron gas of the silicon inversion layer (the very narrow conducting layer at the surface of the silicon in the MOS transistor) by the pioneers of the field, Frank Fang and Alan Fowler of IBM. The unique features of the inversion layer are that electron transport is two-dimensional and that the carrier concentration and Fermi energy can be varied by simply altering the gate voltage. These authors had found a temperature-dependent conductivity with the activation energy decreasing as the number of carriers increased. After studying the paper, he then proposed that this was a model system for investigating Anderson localisation and drew on the board how the very low temperature resistance should behave; essentially the conductivity in the localised region for a particular value of carrier concentration should pass from excitation to extended states to variable-range hopping as the temperature decreased. It was the ability to change the carrier concentration by a wide range and pass through the metal–insulator transition which had seized his imagination. He published his suggestion as part of his 1973 IEEE Kelvin Lecture. Then at Caswell with my colleague Ray Oakley, I was using the MNOS memory device as a system in which the charge at the interface could be varied, with a consequent change in the mobility and binding to the interface charges. However, the difficulty of not having suitable variable-temperature equipment meant that progress was slow. Nevill and the Caswell management arranged for me to take a sabbatical year to work with John Adkins in the Low Temperature Physics group of the Cavendish from September 1973, where the excellent low temperature facilities resulted in rapid progress.

When the sabbatical year had expired, Nevill suggested to the company that the leave of absence should be extended, but by that time the economic situation had deteriorated following the rise in the oil price and the company could offer only unpaid leave. The principal source of any financial support was the Science Research Council (SRC) and this was not easy as my position appeared anomalous (now there are many schemes for the support of people moving between industry and academia). Besides that, the use of inversion layers was uncommon in solid-state physics and not fully appreciated. As a result of his strong support, I managed to obtain a number of short-term grants. I recall the time when a grant application to continue the work had been turned down by the SRC, causing us considerable depression. Shortly after receiving the letter informing us of this, he found himself in London with a spare hour and decided to go to their offices, then in Holborn, so he could find out for himself what the problem was and put it right! One can imagine the consternation which his unannounced and unanticipated arrival created in the office. The application was funded during the next grant round, following which I was fortunate enough to be awarded a Royal Society Research Fellowship.

During those years, we used the silicon MOS transistor as a model system for the investigation of disorder and localisation effects, as well as using the existence of localisation to obtain information on the charge in the oxide, possibly related to his ideas on negative U centres. He greatly appreciated and derived much pleasure from the family connection with the approach of this work, as his father, C. F. Mott, was the first person to attempt investigation of the field-effect device, the precursor of the MOS device. This arose in 1901, shortly after the discovery of the electron in 1897, when J. J. Thomson made a suggestion to C. F. Mott, who was then pursuing research in the Cavendish and is on photographs of the research students during that period. Thomson suggested that if a voltage is applied to the top layer of a metal–insulator–metal sandwich structure, then the electron concentration in the bottom layer is altered, producing a change in electrical resistance. As we now know, the change in electron concentration in a metal is immeasurably small and the consequent effect on resistance is not observable, so unfortunately this line of research was not fruitful. It took the discovery of semiconductors, and the removal of surface states, before the effect could be explored in the 1950s and ever since. Nevill had kept C. F. Mott's notes and results, and I referred to them in a review article.

During the period 1973 to the early 1980s the work using the two-dimensional electron gas in the inversion layer of the silicon device allowed us to explore variable-range hopping, the nature of wave function decay and numerous consequences of disorder, a programme which was extended to gallium arsenide by exploiting impurity band conduction. We also investigated localisation in the presence of a quantising magnetic field, which led to the collaborative project that found the quantum Hall effect. It was a very exciting time as the entire field of semiconductor structures for the pursuit of basic physics was developing rapidly. At all stages he took a keen interest in progress, making suggestions and putting forward new ideas as soon as they occurred to him, frequently discussing the latest results in the laboratory as the experiments were being performed.

We had thought at an early stage that his concept of the minimum metallic conductivity was of universal validity, as so many different experimental systems gave agreement with his predicted value, but subsequent theoretical and experimental work established that the result was produced by phase randomisation and at low temperatures the concept broke down. However, it remains a very useful working hypothesis, and may be relevant to magnetic localisation. It is often observed in two-dimensional systems when strong interaction dominate.

The classic example of his approach is provided by a paper in 1961, where he discusses Bloch–Wilson band theory in which a lattice of monovalent centres will display metallic conduction no matter how great their separation. He states, 'Now for large values of separation this is against common experience and, one might say, common sense.' And indeed, this was how it was with him, as he patiently explained his latest insight to an audience, which could be a research student at a lab bench or several thousand at an international conference, one realised that what was surprising to us was really just scientific common sense to him!

From the mid 1980s he became fascinated by high-T_c superconductivity, whereas I was attempting to follow the consequences of lower dimensionality by fabricating semiconductor structures in which electrons could be electrostatically squeezed to produce a transition to a controlled one-dimensionality and eventually zero dimensions, i.e. total confinement. Even though he was no longer actively working on this topic, he retained an interest in the outcome, especially the consequences of removing the role of disorder. In one dimension there appears the quantised ballistic resistance and in zero dimensions there can appear various single-electron transport effects. As pointed out recently by Philip Anderson, two of the most intensively studied topics in solid-state physics, the quantum Hall effect and systems of dimensionality lower than 2, are descended from interest in localisation, the topic which Nevill Mott recognised in the late 1940s as being of prime importance.

He had often given his opinion that developments in solid-state physics depended on the progress of the relevant technology. Many times he suggested that I should drop a line to his old friend, Sir William (Bill) Penney, who with Kronig proposed the Kronig–Penney model, which is now part of the foundation and history of solid-state physics, to let him know that a model which in the 1930s he considered artificial and purely theoretical was now experimentally accessible and indeed routinely studied. Unhappily, I didn't get around to this before Penney died.

Nevill had a characteristic posture when listening to something which interested him; one leg crossed over the other, handkerchief in hand, holding his chin, he listened carefully, slowly nodding his head. When he thought he appreciated the position he smiled, or often beamed when one of his ideas or theories had appeared in a new or unexpected manner; then he changed the subject in a delightful way. One knew that subsequently there would be many discussions on this and related matters, and if he was away then there would be a considerable correspondence from his cottage in Norfolk, or from a conference, or really from anywhere which possessed a postal system. Frequently, while I was puzzling as to how to answer one letter proposing a new explanation or theory, a further letter would arrive changing his proposal, a process repeated until he was happy that he had arrived at the truth of the matter.

My last discussion with him took place two weeks before he died. We had performed an experiment in which a travelling, surface acoustic wave was established in gallium arsenide. Electrons were trapped in the potential minima of the wave and were carried through the crystal at sound velocity. By passing the wave and associated electrons through a one-dimensional structure, in which a strong confining potential was applied, it was possible progressively to reduce the number of electrons in each minimum to one. This resulted in a current given by $I = fe$, where f is the frequency and e is the electron charge. As f can be several gigahertz, the technique may have eventual application in the measurement of e. However, an intriguing aspect of the experiment is that if one electron is localised in each potential well, and a second is prevented from being present by the electron–electron interaction, then the electrons quite simply form a sliding Mott insulator. I went through this argument and he nodded and listened, clarifying many points as I went

Nevill congratulating Michael Pepper in 1985 after his delivery of the inaugural Mott lecture at the Institute of Physics Condensed Matter Conference, University of Reading.

along. Eventually, I concluded with the thought that a sliding Mott insulator might perhaps allow accurate measurement of the charge on the electron. He nodded slowly, then after a few seconds, he beamed and said, 'Family well?' Sadly there was to be no opportunity for further discussion.

Collaboration with Nevill Mott was unique, his extraordinary insight, interest in experiment, unbounded creativity and passion for new science are virtually impossible to describe. But above all there was the deep friendship which accompanied the science. Meeting him changed the course of my career and for this, and the numerous scientific insights he generously shared, I will always be grateful.

PROFESSOR M. PEPPER, FRS, is Head of Semiconductor Physics in the Cavendish Laboratory, Cambridge where he works on low-dimensional electron transport phenomena. His collaboration with Sir Nevill started in 1971, when he was working at Plessey, and continued throughout the remaining twenty-five years of Nevill's life.

Reminiscences

Lionel and Pearl Friedman

I first met Nevill Mott in 1966, at the Sheffield conference on low mobility solids. My earlier work with Ted Holstein was on small-polaron hopping and the Hall effect due to small-polaron motion. At that time, Nevill was just beginning to explore conduction in the amorphous semiconductors and glasses, so he was very interested in the role of small polarons in these low mobility solids. We exchanged a few scientific comments at that time.

Our next meeting was when I visited the (old) Cavendish on a British Science Research Council Fellowship in 1970–71. I worked on transport and the Hall effect in amorphous semiconductors. Nevill was very friendly and hospitable to me and to my wife, Pearl. We had much pleasurable social contact and he obtained a visiting membership for me at Gonville and Caius College. At that time he consulted for Plessey, the electronics company, and was kind enough to pick me up in his car to join him there for scientific discussions and workshops. On another occasion, Nevill and Ruth invited our family for Christmas dinner. We had a lovely dinner and met his daughter and sister. After dinner, Ruth played their piano and we sang carols. Ruth and Nevill were always fun to be with and they made us feel at home.

Later, when I had a visiting position at the University of East Anglia, and subsequently a faculty position at the University of St Andrews, Scotland, he always made a point of visiting and discussing physics. On one such visit, we enjoyed the Scottish Highland games in nearby Cupar. On another occasion, Nevill, ourselves and other guests enjoyed a long dinner. Nevill had a reservation at a small hotel nearby. But when we arrived, after 10 P.M., it was quite shut. No amount of pounding on the door or telephoning elicited any response. So, we managed to put him up in our home.

Nevill was always very friendly and outgoing. He enjoyed interacting with people, stimulating them scientifically, leading to good and interesting work. He distilled important theory into key ideas and concepts that gave great stimulus to whatever field he was working in. When some physics bothered him, he did not hesitate to organize a short conference at the Cavendish with experts from all over the United Kingdom. These small meetings were very stimulating and interesting. On a lighter note, I remember how, in fun, he enjoyed getting responses in as many foreign languages as possible at international conferences.

Pearl and I once visited him and Lady Mott at their cottage in Blakeney, by the sea. We took a long walk on the sands only to be caught in torrential rain. I had to borrow some dry socks and a jumper from Nevill, which I dutifly returned to him at the Cavendish the next day.

We visited him in his retirement in Milton Keynes on two or so occasions. He met us at the train with his car, took us to lunch, and on one occasion we spoke of the memento he was writing for Herbert Fröhlich, who had just died, and other matters. Finally, he was kind enough to invite me to the workshop on polarons and bipolarons in high-T_c materials, to learn of his latest ideas on this, his last labor of love. It was our privilege to have crossed paths with Nevill Mott.

DR L. FRIEDMAN did his graduate research with Ted Holstein and was on the technical staff at the RCA David Sarnoff Research Center (1963–1970) before visiting the Cavendish Laboratory on an SRC Fellowship (1970–1972). He then held a visiting position at East Anglia University, served on the faculty of St Andrews University (1973–1980) and is currently at US Air Force Rome Laboratory, Massachusetts.

Democritus of Elysium

Mike Pollak

September 16, 1969, following a flight across a continent and over the ocean with two small children, my wife and I arrived for my first visit in Cambridge. It was early evening when we reached our residence at 285 Hills Road. Wearily we attempted to settle into the flat for the night, in the process converting the immaculate flat into a bedlam of rummaged suitcases and crying children. At the peak of the chaos the doorbell rang. My wife, Rosemarie, opened the door. Puzzled, she reported to me that a man she didn't know asked for me. No less puzzled, I went to the door. It took me a few moments, in the dusk, to recognize that the man in rolled-up shirtsleeves, with a warm smile on his face and a twinkle in his eye, was my host, Professor Sir Nevill Mott. The chaos within didn't disturb him in the least. He wanted only to know if we needed help in any way. It was because of his earlier personal efforts that we were able to go directly to a comfortable flat and begin our residence in Cambridge with so little difficulty. Long before our arrival, Nevill had set in motion a pursuit for housing, working with his secretary, Miss Shirley Fieldhouse, and with Mrs Burkill, then President of the Society for Visiting Scholars and his long-time friend.

My wife and I were both touched by this welcoming gesture. To me Professor Mott was an Olympian of Science, far too busy to concern himself with the needs of a young visiting physicist and his family. Rosemarie was awed. She had been nervous about meeting someone knighted by the Queen, worried that she wouldn't know how to behave in his presence. His appearance on our doorstep, so informally dressed and so concerned about our welfare, shattered forever her stereotype of stuffy nobility, replete with monocle and umbrella. Nevill and Rosemarie became good friends and it was her special delight, years later, to host a reception for him in our home when he visited me at the University of California in Riverside.

My personal contact with Mott began late in 1966 when he wrote me, requesting my opinions about the Hall effect and the thermoelectric power associated with hopping. I was then fully involved in biophysics research on some physical properties of DNA which I thought might have some biological significance. But Mott's letter was to be the beginning of a continued correspondence with him, broadening gradually to other topics. We corresponded at some length about the dynamics of hopping in Anderson-localized states versus the dynamics of polaron hopping, about the behavior of AC conductivity at extremely low temperatures, and about variable-range hopping.

Mott's questions and comments were always to the point and were easily understandable. I also noticed very early his attempts to test theoretical results with as many experiments as possible. This appealed to me since I have done both experimental and theoretical work.

The ease of communication with Mott was a great pleasure and a refreshing change to the frustrations I was experiencing in biophysics. In those days, communication with scientists in other specializations was very difficult. I could not resist the attraction of the easy and stimulating interaction with Mott, and drifted gradually back from biophysics into transport in Anderson-localized systems, an area in which I have previously worked. The process was accelerated by Mott's invitation to come for some months to Cambridge, and to write a book on hopping transport. I went to Cambridge, but never managed to complete the book. At the time I was suffering from recurrent depression and the sustained effort needed for the book proved too much under those circumstances. Nevill was very supportive during this difficult period, always expressing sympathy, never resentment over a broken commitment.

The opportunity presented by Nevill's invitation to work at Cavendish opened up a new world – Cambridge was Elysium. I suddenly had the stimulation of extensive discussions with him as well as with some of his associates, notably Mike Pepper, Ted Davis, John Adkins, Abe Yoffe and Yao Liang. It was a far cry from home base, where my scientific interaction was largely limited to my graduate students (one of whom I brought with me to Cambridge). Mott, at the time, was Cavendish Professor, busy doing all that this most prestigious post entails. There seems to be consensus that in the distinguished succession of Cavendish Professors he was among the best. Yet he was readily available for discussion and I happily availed myself of this opportunity. Our talks ranged over a variety of subjects, always resulting in my deeper understanding of the problem at hand. It seems to me that Mott's own quick and deep insight into physical problems (even those he'd never considered before) enabled him to focus precisely on the heart of the subject.

My future work was greatly influenced by discussions we had in Cambridge on the role of interactions, and of percolation, in transport in Anderson insulators. The possibility to describe hopping conduction by percolation had occurred to me when, during the course of some correspondence on his now famous variable-range hopping conduction, Mott asked me if his theory could be put on a firmer footing. It so happened that percolation was connected with some earlier work I did on low frequency hopping conduction.

Having been both encouraged and aided by discussions with Mott, I worked on interactions and on percolation after my first visit to Cambridge and consider them to be among my more significant works. I presented my first paper on interactions at a conference in Bristol in 1970 and on percolative hopping transport at a conference in Ann Arbor in 1971.

Work with Mott, always stimulating and productive, was sufficient reason to value a stay in Cambridge. But Nevill also took care to see that visitors had many enjoyable personal experiences. There were invitations to Ruth and Nevill's home, to sherry parties, to special Christmas events organized for the children and to pleasant afternoon teas with his daughter and her family. A cherished memory is Nevill sitting on the floor with our daughters, then ages 2 and 4, playing with them and some wooden toy blocks he had kept from his own childhood. Unforgettable though it is, I wish I had had a camera. Our daughters, Michelle and Tania, would surely have treasured a photograph.

My contact with Mott continued for many years after leaving Cambridge. There were repeated visits to Cambridge, including another long-term visit in 1977, meetings at conferences, Mott's visit to Riverside, and a long-term correspondence. I appreciated Nevill's interest in my opinion about some of his work as well as other work in the field. The most extensive discussions continued to be on the Coulomb problem in Anderson insulators, particularly the effect of interactions on transport. My collaborators (Mike Knotek, Miguel Ortuno and several others) and I wrote a number of papers on the subject while in close communication with Mott. We all agreed that certain many-body effects (similar to when a dancer adapts to the motion of others on the floor) must be important, although neglected in certain well-known theories. However, we never quite came to agree on the more detailed nature of such a correlated motion. At Mott's suggestion we gave back-to-back papers on the subject at a conference in Cambridge, Massachusetts, in 1979 and afterwards had some intensive discussions in private. Mott came closest to expressing agreement with my view in his closing talk at that conference.

At one point we planned to write a joint paper on the role of interactions in transport. But we backed away from the plan, partly because of some remaining disagreements and partly because of the difficulty communicating with Efros and Shklovskii in the USSR, who also came to write on the subject. We were in disagreement with their work but felt that without communicating with them we could not write a substantive paper on the problem. Despite wide interest in the electronic transport in random media, the very question of the importance of many-body effects remains controversial. Mott, in his latest writing on the subject,[1] considered the issue as unresolved. Only very recently some more definitive results on the problem could be obtained, mainly thanks to the rapid development of computer capabilities.

The difficulty of Soviet scientists to communicate with the West and to participate in international conferences was a substantial hindrance to the development of certain branches of science and of course an aspect of humanitarian injustice. Both were anathema to Mott and he readily helped in any such situation he could. At one conference I was cautiously approached by a Soviet bloc participant. He had a message from Efros and Shklovskii asking if I could assist with the publication and translation of their book on properties of doped semiconductors. Realizing that such a publication could facilitate their attempt to come to the West, I quickly contacted Mott, also at the conference. He immediately offered his help and promptly contacted a publisher who agreed to publish the book in English. There was a problem finding a suitable translator, but one finally materialized and the book was published. Both scientists ultimately came to the West.

My interaction with Mott took place during the years when his research was focused on non-crystalline solids. I remember that before then I heard him express (at an international conference in Exeter in 1962) some concern that solid-state physics had no new fundamental problems to offer. But a few years later he foresaw the enormous latent interest in disordered systems, with a potentially wide spectrum of fundamental problems and practical applications. I believe there is agreement that he was the driving force responsible for making the subject one of the most active in physics. I think he accomplished that by a sustained effort to encourage the direct involvement of numerous scientists through personal discussions, extensive correspondence and organisation of topical conferences. This is besides his many important contributions which led to a Nobel prize. It is worthwhile noting that even Schrödinger, who already in the forties stressed[2] the extraordinary interest inherent in the 'aperiodic solid', did not succeed in generating wide excitement about the subject.

An incident at a conference in Edinburgh, at the same time as the Edinburgh Festival, is characteristic of Mott's quest for understanding physics and, if possible, socializing in the process. On a free evening when many attendees planned to see the spectacular Tattoo, Mott thought that a certain unresolved problem deserved discussion by a group of participants. Recognizing the attraction of the Tattoo, he proposed a discussion party and promised to provide a bottle of superior malt whisky. The few who elected to attend the Tattoo ended up drenched in a sudden storm. Those of us who joined Mott enjoyed a very pleasant evening of discussions, stories and some truly superb malt.

I have been fortunate and privileged to have enjoyed both a professional and personal relationship with Nevill. It is difficult to attach a definition to the concept of a great scientist, but to me Mott was a modern Democritus. While Democritus two and a half millennia ago pioneered the concept that all materials are built of atoms, and their properties depend on the atomic structure, Mott was a pioneer in applying modern atomic theory to that problem. It is deeply gratifying that he was a personal friend, a relationship that contributed greatly to making my work in science a pleasant experience. I miss him very much.

References

1. MOTT, N. F. (1993) *Conduction in Non-Crystalline Materials*, 2nd edition. Oxford: Clarendon Press.
2. SCHRÖDINGER, E. (1944) *What is Life: The Physical Aspect of the Living Cell*. Cambridge: Cambridge University Press.

PROFESSOR M. POLLAK was born in what was then Czechoslovakia in 1926. Thanks to his parents' foresight he survived the Nazi occupation by emigrating to Palestine. He obtained a BSc in electrical engineering from the Haifa Technion and an MS and PhD in physics from the University of Pittsburgh. After spending time at the Bell Telephone Laboratories and Westinghouse Research Laboratories, he joined the physics department at the University of California, Riverside, in 1966. His primary interests are in the physics of disordered systems.

Defects in Chalcogenide Glasses

Bob Street

Nevill Mott was head of the Cavendish when I started my PhD in 1968 and it should be no surprise that he suggested I study amorphous semiconductors. However, apart from one incident described shortly, my main interaction with him did not start until a few years later. The end of the 1960s was still early in the research of amorphous semiconductors. The third international conference on amorphous semiconductors had just taken place at Cambridge, naturally organized and run by Nevill; part of a biennial series, it has now reached number seventeen. Ted Davis had come to Cambridge and there was an active experimental group at the Cavendish to complement Nevill's theory. Much of the experimental work in the field was then aimed at understanding the electronic properties of chalcogenide glasses, typified by selenium, As_2Se_3 and similar materials. The puzzles they presented focused on the observation that electrical conductivity measurements showed the Fermi energy was pinned in the middle of the band gap but, unlike the amorphous silicon being studied at the time, spin resonance and hopping conductivity gave little indication of the expected defects.

Nevill's style of theoretical physics kept him in close contact with experimental work, sometimes too close for his own good. My first contact with him concerning conductivity in chalcogenides, quite early on during my PhD work, was an amusing revelation about scientific research. It had been known and understood for several years that hopping conductivity had a certain frequency dependence. Just before I started at Cambridge, a different dependence had been observed at higher frequency. Nevill was developing some creative theories to explain the behavior and I started further measurements of the effect. Alas, it turned out to be an experimental artifact completely unrelated to the material. Some nice theory had to be abandoned and we all turned to other problems.

Our real collaboration began a few years later, again through Nevill's astonishing ability to connect theory and experiment. He and Ted Davis were trying to come up with a comprehensive model for defect states in chalcogenides, one that would explain all the electrical and optical data. By this time, 1973-74, I had left Cambridge for a postdoctoral position at Sheffield, with Ifor Austin and Tim Searle, where I was studying the luminescence of the chalcogenide glasses. In the course of this work there was rapidly solidifying evidence that the emission was from charged defects and, more significant, there was a clear indication for lattice relaxation. The Stokes shift separating the emission and absorption energy was the telltale sign.

I'm not sure how Mott heard of this result, but it was probably at one of the Chelsea meetings. These were the annual get-togethers of the UK Amorphous Semiconductor Group, meeting at Chelsea College just before Christmas each year. It was a great chance to meet old friends, present new work in an informal atmosphere, usually with Nevill in charge, and of course for an early celebration of Christmas.

Around the same time, Phil Anderson proposed an explanation of the chalcogenide glass puzzle using the new concept of negative correlation energy. This was a mechanism for pairing electronic states that allowed no spins and a pinned Fermi energy – just what was needed to explain the data. Furthermore, atomic relaxation was the essential feature which allowed the states to gain enough energy to overcome the electron repulsion. However, Nevill didn't believe the model. Anderson applied the negative-U concept to the entire material, whereas there was a growing body of evidence that the properties of chalcogenides were indeed controlled by defects. Since Anderson's theory was characteristically difficult, Nevill was probably the only person who understood in depth both the theory and the experimental evidence for defects.

The negative-U idea did provide a neat solution to the spin problem. If defects in As_2Se_3 had lattice relaxation and negative-U states required lattice relaxation, then negative-U defects seemed the plausible explanation. This sparked a collaboration between Nevill, Ted and myself to develop the model and to show that it did indeed explain the electronic properties. Eventually this work exceeded our expectations as it not only provided a detailed model of virtually all the defect-related electronic properties, but also had a simple explanation of why the defect behaved in this way, based on the chemical bonding. It took over two years during 1974–76 to work all this out, and continuing ideas and repercussions followed for several years afterwards. What an opportunity it was for myself near the start of my career to work so closely with a scientist such as Nevill Mott.

The collaboration was close but the distance was not. During this time I left Sheffield for the Max Planck Institute in Stuttgart. Apart from a few occasions when we could meet, the entire collaboration was done by letter. These were traded back and forth at a brisk pace, often two a week and varying from very short comments to long analysis. Reading his letters again reminded me how much to the point they were – almost like finished papers. It also recalled the days before e.mail.

We started out trying to fit together the ideas of charged defects and lattice relaxation, eventually using the negative-U idea to come up with a defect reaction to describe the electronic states and their transitions: $D^0 \rightarrow D^+ + D^-$. A good deal of effort went into the analysis of energy levels to show how this approach explained conductivity, transport, optical absorption and luminescence. In particular, the metastable D^0 state nicely explained the induced electron spin resonance (ESR) and absorption. What we called the MDS model was published in 1975.

Two things stand out from this collaboration, one of which was obvious at the time and the other only by later reflection. Nevill immediately made it a very equal relationship, despite his having been doing research for forty years to my four (to say nothing of his stature in the field). I learned quickly that this was his style, but initially it was both surprising and gratifying that he gave so much attention to my views on his ideas.

Needless to say, I devoted a lot of time to this work, since merely keeping up my end of the flow of letters was quite a challenge. At the time I simply accepted that Nevill was similarly engaged, evidently ready to think through a point and reply as soon as a letter arrived. Indeed, as we went along, he regularly wrote three or four page notes and comments on various aspects of the problem, and they were circulated quite widely. In hindsight I can only admire that he had the time and energy, since he was obviously involved in all manner of activities within Cambridge and at a national level. He was undoubtedly a man of great energy.

We met a few times during this period in Paris, Marburg and of course at the Chelsea meetings. In 1974 we arranged to meet Josef Stuke in Marburg. Nevill and I met at Frankfurt station to complete the last part of the train journey. No sooner had we sat down in the compartment than he opened up his small gray suitcase, rummaged around under the pyjamas, pulled out some papers and started to discuss physics. The discussion continued for the next couple of days.

The collaboration continued after I moved to Palo Alto in 1976, although more intermittently. We applied the model to the effects of impurities and to SiO_2. Much experimental work, particularly on metastable effects, was shown to fit nicely into the model. My interests changed to amorphous silicon, in which surprisingly similar ideas of defect reactions, but without the negative U, turned out to be important in describing the doping.

DR R. A. STREET studied physics at Cambridge University, receiving his BA degree in 1968 and PhD in 1971. After a postdoctoral fellowship at Sheffield University, he became a visiting scientist at the Max Planck Institute in Stuttgart. Since 1976 he has been at the Xerox Palo Alto Research Center and is presently a Senior Research Fellow and Area Manager of the Imaging Systems and Materials Group. His research has concentrated on the material and device properties of hydrogenated amorphous silicon.

Glass, Raspberries and Cader Idris

Neville Greaves

Nevill Mott shines through as one of the greatest of physicists, judged both in world terms and in terms of this century. He has set condensed matter physics on its feet and his prolific output has added gloriously to the supreme legacy of European science. At the same time, he has been the touchstone for many of us caught up in the mysterious zone between conceptual idea and experimental fact. Approachability and genius are not often found together but they coincided in Nevill Mott; he advanced a great deal of science but he also inspired many affectionate memories.

It can be no offence to say that Nevill was a wonderful eccentric in the best British tradition. I well remember the third conference on amorphous and liquid semiconductors held in Cambridge during 1969. This was the first international conference I had attended and the first time I had encountered Nevill Mott or a session chairman smoking from behind a No Smoking notice. It was the same person who got to his feet following a paper denying the concept of an energy gap in the amorphous state. Looking through a glass beaker, held aloft, and shaking his head slightly as he always did, he began, 'This is undeniably a glass, but it seems to me it has a perfectly good energy gap!' Nevill Mott was a past master at ignoring conventions at any level and somehow touching the heart of the matter.

The Cambridge conference was where I met Abe Yoffe, and through him I came from R&D in the glass industry to a PhD at the Cavendish. At that time Nevill Mott had an office on Free School Lane, where I nervously introduced myself and put forward some ideas I had on small polarons. Right from the start, as others will attest, he was interested, courteous and always provided advice to move ideas forward. On this occasion he suggested I should talk to Lionel Friedmann, which proved just right.

Nevill was a regular attendee at the weekly seminars on physics and chemistry of solids held in Abe's office upstairs. He would sit or recline at the front in an armchair near the window, handkerchief to his mouth, and wait to hear anything new. He generally had something helpful to say at the end. There was one occasion, however, when a certain visiting scientist from the States, whom we called Herb, after talking on dielectric properties for an hour and anxious to get to the Mott question, rather tactlessly dropped all the main titles and asked, 'Well, Dr Mott, what do you think?' At which point the aquiline Sir Nevill rose from the easy chair, put his handkerchief back in his pocket and left the room to a shocked silence but with a very affirmative answer.

It was also about this time that Mott and Davis was published. After the seminal *Phil. Mag.* series of papers, which I had pored over in Pilkington's research laboratories before coming to Cambridge, the book seemed to appear as if by magic – full of so much experimental detail and with all the concepts linked up and plausible. It was reassuring not just to have a text on non-crystalline materials for the first time, but also a practical 'street map' to link experiment to theoretical concept. Ted Davis and I were then working on amorphous arsenic and its crystalline allotropes. Among many other things, we observed how the electrical conductivity, whether as a wide-gap semiconductor, as a narrow-gap semiconductor or as a semimetal, all beautifully approached Mott's minimum metallic conductivity as the highest temperatures were reached. I remember excitedly showing Nevill the multiple plot in Ted's office at the new Mott Building on Madingley Road. 'Yes, but I would have been very surprised if they hadn't,' was the comment that came back. Although I was initially disappointed, it underscored the fact that surprises for him were more important than confirmation, once he felt that a particular point was settled.

The surprises in non-crystalline materials which naturally followed came in the area of defects in glasses, indeed in understanding the electrical and optical properties of silica. By that time I was back in industry and working on the transparency of optical fibres. In putting my ideas together, I benefited from many letters and conversations with Nevill, not least on Bob Week's paramagnetic E centre. 'An oxygen vacancy that traps an electron is fine but where's the hole?' was my question when I visited him one weekend. 'That's exactly my problem!' came the reply. Paired defects were the answer, but with charge virement between cations and anions. He took me round the Cavendish library, hunting for clues.

Before I left Pilkington to join Daresbury Laboratory I received a curious invitation. He was lecturing in Liverpool in February 1978 and would it be convenient for me join him at the Adelphi Hotel at three o'clock for discussions? This proved to be an unusual encounter. I found him halfway down the impressive if gloomy colonial lounge with afternoon tea spread out in front of him. He was concluding a meeting on education with someone from C. F. Mott College. As I drew up my chair, he began, 'So you have a new job. I don't suppose, though, that as a research scientist you will be earning more than a miner.' Since coal miners' salaries were topical at that time, I had to admit that he was right, but I was touched by his concern. We talked about what I planned to do next and then moved off to Oliver Lodge for his seminar. Afterwards his hosts and I went with him for dinner at a large Chinese restaurant about a mile from the university. Although the meal was excellent, the conversation was less so, and his attention wandered to a lively party on an adjoining table. 'And who are the gals?' he suddenly asked. His hosts had to admit they didn't know. 'Pity,' came the reply. Table talk came to an abrupt halt!

Nevill Mott visited Daresbury on several occasions. The first was in 1981 when the synchrotron radiation source was being commissioned and the initial experiments were beginning to roll from the EXAFS team that I had established. It was in the summertime and afterwards he wrote, 'Thank you for an interesting visit and please thank Mrs Greaves for a lovely dinner party – and I did enjoy picking raspberries with your family!' The dinner party was noteworthy on account of John Pendry discovering a plastic engine driver at the bottom of his wine goblet. The children were of course blamed, but at the fruit farm earlier in the evening they had been trying to fathom the justice of why they were not allowed to eat as they picked whereas certain adults were!

Two years before he died, Nevill lectured in Daresbury in the spring and at a European conference that I organised at Chester on synchrotron radiation in materials science in the summer. Both visits were memorable. High T_c superconductivity was the common theme but the human element was very much to the fore. Here was a person full of ideas and speculations and at the same time almost dismissive of his previous work. Some of my colleagues at Daresbury were working on spin-polarised photoemission and were anxious to show him their large Mott detector. Just six months before his ninetieth birthday he clambered up a gantry at the synchrotron radiation source to see the engineering outcome of a problem in the scattering of relativistic electrons he'd solved in his twenties: 'I'd almost forgotten about that but I'm glad it's proved to be useful.' We talked that evening about many things outside the sphere of physics. There was the incompatibility of pursuing free trade and retaining sovereignty: 'You can't have both!' Concerning his Christian beliefs and his commitment to the Anglican persuasion he said, 'When so many people have experienced this for so long, they can't all be wrong.' He seemed to maintain an air of pragmatism in everything he thought about.

The impression he made at Chester was profound. He was led up the steps of the rostrum holding the hand of Gerhard Zachman from the University of Hamburg. He then proceeded to talk for forty minutes without notes about his work with Alexandrov. Not unexpectedly, he did flourish from time to time one of his now legendary handwritten viewfoils, replete with the odd thermopower formula and associated experimental trends, but his argument was always direct and succinct. The familiar pauses in delivery were there, but as usual they punctuated what were otherwise perfectly complete sentences. The following evening he addressed the conference in the splendour of Cheshire's Arley Hall. He recounted, as he had done on several recent occasions, how as a young scientist he had moved between the mathematical courts of Europe. He remembered the exodus of Jewish physicists from Nazi Germany before the start of the Second World War. Always conscious of the contemporary question, he drew a parallel with the current plight of Soviet physicists looking for a 'desk in the West' following the end of the Cold War. All of this left his international audience acutely aware of the precious connection with the golden age of quantum mechanics which he represented.

The last letter I received from Nevill was in April 1996, just after I had been appointed to the chair of physics at the University of Wales,

Nevill with Neville Greaves outside Arley Hall, Cheshire, on the occasion of the First European Conference on Synchrotron Radiation, July 1994.

Aberystwyth. 'Many congratulations,' he wrote. 'I have never visited Aberystwyth – though in my youth I have walked over Cader Idris and those parts. I should love to visit you when installed.' Sadly this was never to be. All I can say, alongside several generations of grateful physicists, is that his fundamental ideas, his encouragement and especially his friendly spirit remain all-pervading. Byw yw ysbrid yr ysbryddwr! (The spirit of the inspirer lives on!).

PROFESSOR G. N. GREAVES graduated from St Andrews then worked in industry; during this time he obtained his PhD at the Cavendish. In 1978 he joined the staff of the synchrotron radiation source at Daresbury Laboratory, becoming Head of Materials Science in 1990. He was appointed to the chair of physics at the University of Wales, Aberystwyth, in 1996. His research interests span the structure and electrical properties of glasses, amorphous semiconductors and ceramics and also include the development of X-ray synchrotron radiation techniques.

Memories of a Graduate Student

Eugenia Mytilineou

My memories of Professor Mott? Without exaggeration, they can be summarised in a few words: he changed my life! His reference letter to Kings College gave me a grant and thus the opportunity of studying for a PhD at the University of Cambridge. There is no doubt that my life would be very different if I had stayed in Greece. He was and still is my hero. Now, as an Assistant Professor in the University of Patras, I talk to my students very often about him as an example of a really great man but still the most modest person I have met.

How did I meet him? In 1974 I had just started working for my PhD at the University of Athens, with Professor M. Roilos, on the Hall effect of the chalcogenide glasses. Somehow I thought that it could be more interesting if I spent my two-years' savings and my summer holidays in the Cavendish Laboratory instead of a Greek island! It was the time of a great dispute about the sign reversal between the thermopower and the Hall effect in the chalcogenide glasses, and Professor E. A. (Ted) Davis accepted my proposal with pleasure. On my arrival, Ted found himself confronted by two surprises, which in spite of his perfect English manners, he could not hide: I was a woman and not as senior as he had thought! But nothing stopped him from eventually becoming my PhD supervisor and we maintain a perfect relationship and collaboration up to now!

As it was the first time I had left home, I was very shy; my English was very poor and I felt a bit lost. One day, Ted introduced me to Professor Mott, in the cafeteria of the Cavendish Laboratory. Suddenly, I found myself in front of the 'holy master' of amorphous materials and probably of the whole of solid-state physics; almost every chapter in any textbook contains a law named after him! He stood up in welcome, and as he was so much taller than me, I felt as though I was standing next to a giant with a very kind smile. He looked at the floor, not straight at me, and surprisingly he seemed more shy than I was! His smile immediately made me feel very comfortable, even if I didn't understand a word of what he told me. At that time, I could understand only Ted's English, so he had to repeat what other people were saying.

The summer of '74 not only made a mark in my life, but it happened to be a mark in Greek and Cypriot history as well. After a coup d'état in Cyprus, the Turks invaded the island and Greece declared war on Turkey. The borders were closed and I found myself stuck in Cambridge! I stayed a month longer than I originally planned and was getting short of money. The problem was solved by calling the presentation of my results an invited talk, in a small meeting that was organised in the Cavendish. I will not forget my surprise when Professor Mott gave me a cheque from his own personal account, saying in his shy, smiling way, 'The procedure of claims is very slow, you may need the money.' I returned to Greece after democracy was established and life returned to normal. Next year I returned to the Cavendish as a PhD student.

I was fascinated by the warmth, the support and the real care of everyone around. I was also taught a very precious lesson: a professor is not an emperor that nobody can reach; he can be close to the students and still be respected. Greek reality was so different at that time. You could never see or talk to the professor himself, you had always to deal with the assistants! Professor Mott was the person I heard say many times, 'Oh, I don't know!' or 'I don't understand maths!' I also remembered Abe Yoffe's remark, 'He is of an age and authority at which he is not afraid of anybody not believing him!' Since then, I try (at least) to say more often, 'I don't know,' trying hard to overcome my fears.

As the years followed, we had a lot of friendly discussions, not only on physics but on Ancient Greek history. With his wide and multifaceted personality he influenced my way of thinking and motivated me to go back and read about Ancient Greek civilisation and Byzantine painting! Once, when I was in Greece, I sent him a postcard of *Spring*, a marvellous fresco from Santorini (Thera). A little island in the middle of the Aegean Sea, it suffered a volcanic eruption in 1200 BC; half of the island blew up and covered the town with three metres of ash. A very important commercial and cultural society vanished from the face of earth. Some people like to identify it with the lost continent of Atlantis. The town was located and excavated around 1970. Professor Mott was not aware of this and he became very interested. When I returned from my holidays, he came to my desk and started asking me all sort of questions. If anybody came to talk to him, he asked politely, 'Could you please come back later, we talk about Greek history now!' It was as if facing a new physical problem, he wanted it analysed down to first principles, the basic laws. Very often I did not know what to answer or I had to think hard to get a satisfactory explanation. One of the questions I recall was, 'Did the eruption happen before or after the Trojan War?' What was that, a comparative and critical study of history? I never learned it in school! How limited my education was!

Another time, he returned from a conference in Rome. Very excited, he came to talk to me about the Byzantine churches he visited in Rome. He said, 'People usually visit ancient Rome and are not aware that there are very important Byzantine churches there.' He made me realise that the Byzantine icons in Italy are very important for studying the transformation of the two-dimensional Byzantine Virgin Marys to the three-dimensional Madonnas with Child and the birth of the Renaissance!

On another occasion he asked me about Greek spelling. In the Greek language the same sound corresponds to different symbols (letters or group of letters). I mentioned to him the theory that originally the different letters had a different pronunciation, but in the years of Alexander the Great, when the Greek language spread from the Mediterranean up to India, the proper pronunciation was lost, but the spelling remained. He was quiet for a while and then, 'Oh! yes, it is a good approach. I understand, it is exactly what the Americans did to the English language!'

The last time I met him was at the Cambridge ICAS conference in 1993. He had a stick and looked old but still with a critical bright mind. Walking back to our rooms, he commented, 'My body left me, I need a stick now.' Ted e.mailed me the news of his death. I was shocked. It was as though I had lost somebody from my family, the person who gave me the opportunity of obtaining a PhD in Cambridge and whose example influenced my character and my way of thinking. I had the opportunity to be at his memorial service. I saw old familiar faces and places. His presence was alive among us. He will always be in my memory.

PROFESSOR EUGENIA (GENIE) MYTILINEOU did her PhD in the University of Cambridge from 1975 to 1978. Now she is an Assistant Professor in the University of Patras, Greece. Her research interests are principally in the electrical and optical properties of amorphous semiconductors.

The Mott Transition

Gordon Thomas

Professor Sir Nevill Mott stood at the centre of the attempt by twentieth-century scientists to understand metals and insulators and how each can change into the other. He emphasised his focus on this area of research in a succinct and modest description of himself for the book that he edited in 1991, *Can Scientists Believe?*:

> To me, working on semiconductors and the like, the practice of science is the unravelling of one small puzzle after another, why some insulating materials start to conduct electricity when you compress them, why some complicated compounds lose their electrical resistance at low temperatures.

Mott not only unravelled some of the puzzles of his field, but he also led the field in formulating the problems in ways that seemed elegant and therefore appealing to many scientists in the past thirty years, including myself. Nevertheless, to other scientists, who perhaps had not yet studied Mott's books and papers, particularly his 1974 book, *The Metal-Insulator Transition*, the question of the essence of metallic and insulating states seemed complicated, perhaps less than fundamental, and therefore something to be deferred. For instance, Richard Feynman alluded to the subject with a bit of this tone in his 1965 *Lectures on Physics*:

> Some materials are electrical 'conductors' because their electrons are free to move about; others are insulators – because their electrons are held tightly to individual atoms. We shall consider later how some of these properties come about, but that is a very complicated matter.

Yet Mott was undeterred and even enthusiastic about a search for order in the apparent confusion of innumerable electrons and atoms interacting with each other. One who shared Mott's interest was Philip W. Anderson, who strongly argued for the importance of such difficult aspects of nature, as he explained in a 1972 essay 'More is Different':

> We expect to encounter fascinating and, I believe, very fundamental questions at each stage in fitting together less complicated pieces into the more complicated system and understanding the basically new types of behaviour that can result.

The metal–insulator is certainly an important example of this type of complex, but fundamental, scientific question. Its importance and Mott's contributions to it over at least a third of a century were recognised by the Swedish Academy of Sciences in 1977 with a Nobel prize (shared with Anderson and J. H. Van Vleck). My personal high opinion of Mott's theoretical work is indicated by the fact that it inspired a series of my own experimental investigations over the past twenty-five years.

Besides being a scientific pioneer, Mott was also remarkably open to new ideas, even ideas that differed substantially from what he had previously thought. For example, Mott forged a friendship with me even after my colleagues and I failed to find the metal–insulator transition in the form he had predicted. That open-mindedness was also a sign of his greatness.

To illustrate my relationship with Mott, let me recount some recent events. On 3 August 1996, Mott was very much on my mind as I rose to give the opening address at the International Conference on Electron Localisation and Quantum Transport in Solids, held in Jaszowiec, Poland. Out of the window of the small conference hotel, I saw a light rain falling through a mist on the darkly forested hillside. Inside the room I saw people filling every chair and standing at the back, but Mott was absent. On my chair, my copy of the programme still listed him as the plenary speaker at the midpoint of the conference. I projected my title, 'Search for the Mott Transition: A Tribute to Sir Nevill', and opened my talk by saying, 'I believe that Professor Mott is now at home in his house in Aspley Guise, probably having tea. He wanted to be here.' I did not know then that Mott was fatally ill and would die within a few days, but I did know that, despite his age of more than ninety years, he wanted to be at the conference. Exactly a year earlier, on 3 August 1995, he had written me from his home, asking me to travel with him to Poland for this conference. We made plans for the trip, and he wrote, referring to his research plans, 'I want to translate all that is known about the Mott–Anderson transition to the case when the carriers are bosons.' Just before the conference, I called him at home for a final confirmation of our travel plans. He answered the phone himself but said, with clear regret, that his doctor had just forbidden him to make the trip.

Since Mott could not give his own talk, I modified mine to present an overview of how his work had influenced our knowledge of metals and insulators. In particular, I discussed three types of materials: electron–hole liquids, vanadium oxides and doped semiconductors. I explained how many people, including my collaborators and myself, had sought a discontinuous transition – dominated by interactions among the electrons – of the sort he had predicted, but other effects had somehow dominated each of these systems. Nonetheless, I emphasised, Mott had embraced our findings with undiminished enthusiasm.

The first illustration in my talk was the transition among electrons and holes. In the 1970s my colleagues and I had cooled a pure crystal of germanium so that there were essentially no free electrons, and we had then focused a beam of light on the crystal to excite just a few electrons. These liberated electrons leave holes behind on the germanium lattice; although both the electrons and holes move around easily, each electron binds to a hole to form an exciton (sometimes called a Mott exciton since he had described such large and mobile bound charges). As we increased the intensity of the light, we made more excitons and had moved toward the concentration at which Mott had predicted the gas of excitons would turn into a metal (when the spacing between excitons was four times the exciton radius). However, liquid droplets condensed before we could reach the transition. Inside the droplets, the charge density was high enough that we found the electrons and holes to be unbound, and thus metallic, as Mott had predicted. However, the transition itself was inaccessible because of the liquid condensation. We found a distortion in the liquid–gas phase diagram near the Mott density and called it a Mott distortion in a 1978 paper.

I went on to describe our measurements of crystalline vanadium sesquioxide and how they had also failed to show the sort of discontinuous transition that Mott had predicted. With cooling and the application of pressure or the addition of charge carriers, this system behaved as if it were approaching a gradual metal–insulator transition. However, just before the transition, the crystal structure changed drastically and again prevented an observation of Mott's effect in the pure form that he had envisioned. Nevertheless, Mott remained enthusiastic about our results, even though the transition we observed showed a tendency toward being continuous and, in the end, was discontinuous because of the crystal lattice, not because of the interactions among the electrons, which had been Mott's focus.

As I also recounted at the conference, I had worried unnecessarily in 1980 that Mott might be particularly displeased when my colleagues and I reported the absence of his minimum metallic conductivity in doped semiconductors only a few years after his Nobel prize. In this case, instead of excitons, we had put phosphorus atoms into crystalline silicon because (since the phosphorus ions were essentially immobile) a liquid–gas phase separation could not occur during our experiment. Furthermore, the crystal structure of the silicon was stable like that of diamond. We cooled a succession of samples to temperatures only a few thousandths of a degree above absolute zero, and in samples with electron densities (produced by the phosphorus impurities) above which Mott had predicted a metal, we found metallic behaviour. In other words, the electrons continued to be able to move through the whole sample with only a gentle push as the temperature approached absolute zero. The exciting effect we then tried to observe was Mott's prediction that this base conductivity (a measure of the intrinsic ease of the electrons' motion) would decrease with decreasing electron density, but jump suddenly to zero at the transition. We found that, although the minimum metallic conductivity (before the jump) appeared to have been well established in the previous literature, including review articles and Mott's 1974 book, *The Metal–Insulator Transition*, it did not exist in our experiments. Despite the fact that we had disproved an important part of his theory, Mott responded, when I sent him a copy of our 1980 paper, that he found the results very convincing and agreed with us that the random arrangement of the impurities must play a major role, as Anderson had predicted. Instead of attacking our work, he added that he looked forward to the opportunity to develop a new theory and invited me to visit him. In the succeeding years, we exchanged many letters and several visits to both our homes and offices. (Even my parents, my wife and my son came to know him.) He drastically revised the 1990 edition of *The Metal–Insulator Transition* and emphasised the revisions in his new preface.

I ended my 1996 talk there, but research on the metal–insulator transition goes on. The jacket of *Can Scientists Believe?* (1991) asserts that Mott 'was awarded the Nobel prize for work done mainly after his retirement and he remains active in this field.' I am tempted to quibble that he never really retired. Moreover, despite his death, he will continue to be an active influence on the metal–insulator transition because this field will continue to develop from his ideas. As Nathaniel Hawthorn enquired in 1851 in *The House of Seven Gables*:

Is it a fact, or have I dreamed it – that, by means of electricity, the world of matter has become a great nerve, vibrating thousands of miles in a breathless point of time?

Gordon Thomas with Jan Tauc and Nevill (1986).

Today, Hawthorne's idea of the world is no dream. My experience with twentieth-century physics and technology suggests that Mott's work has affected our world of matter in an important way and will continue to do so in the future. In that we influence the world through its memories of our acts and words, the transitions at our death are gradual ones. Mott's death marks such a transition from his generous acts and theoretical discoveries to the yet unknown works of those moving in the directions to which he pointed.

DR G. A. THOMAS has been a member of the technical staff at Bell Laboratories (now a division of Lucent Technologies) since 1972, except for temporary appointments on the faculty at Harvard University and the University of Tokyo. He is a Fellow of the American Physical Society and has published widely in a number of fields within condensed matter physics, including the metal–insulator transition, superconductivity and communications technology.

A Sovereign Friend

Friederich Hensel

It has been more than thirty years since the morning in which one of my colleagues, J. Stuke, appeared in my laboratory in the Technical University in Karlsruhe with a visitor: a tall man with an immense charm and friendliness who was interested in our results on the electrical conductivity of fluid mercury at high temperatures and pressures. Apparently the visitor thought that our results might prove useful in understanding the problem of localization of electrons and the metal–nonmetal transition in disordered systems. This was my introduction to Professor Mott, whom I knew at that time only by name from his classical books, Mott and H. Jones (1936) *Metals and Alloys*, and Mott and R. W. Gurney (1948) *Electronic Processes in Ionic Crystals*. I was very much impressed and surprised that the famous theorist was willing to listen to a young experimenter, and got so excited that I was convinced my English, which had been good enough to read scientific papers, was not adequate for discussion with the great man. I need hardly say that I completely lost these feelings after a few minutes of talking with him. His willingness to understand our results and his obvious knowledge of German (arising from his time as a student of Max Born in Göttingen) were very helpful to keep my presentation of our work going on.

It was incredibly nice of him to listen for a few minutes to what I said about the metal–nonmetal transition in mercury but then the discussion became more and more Mott and less and less me. I tried in particular to draw attention to a plot which revealed an eight orders of magnitude change in the conductivity of mercury with decreasing density, but he was not very much impressed by this 'gigantic' effect. He turned the discussion instead to a small change in the slope of the conductivity versus density curve which occurred at a conductivity of about 200 ohm^{-1} cm^{-1}, and pointed out, 'This represents the minimum metallic conductivity.' I am convinced this was an important moment for my career as a scientist. His strong interest in the concept of the 'minimum metallic conductivity' led him to include our results on mercury in many articles and in his books. In this way our work was recognized by the scientific community. But far more important for me was his inspiration, guidance, enthusiasm and friendship over more than thirty years. I benefited tremendously from keeping up an intense correspondence between us and from many enlightening conversations which we had during meetings at conferences and in particular during his many visits to Marburg. Both had a profound influence on the way I do things today.

I have particularly vivid and delightful memories of social meetings with Nevill. I remember one Saturday afternoon when my wife, Ursula, and I went with him on an excursion (Nevill prefered the German word *Ausflug*) to Alsfeld, an old town with beautiful half-timbered houses and an interesting medieval town hall. Incidentally, Ursula was able to organize a special guided tour for us through the town hall and two fascinating old churches. Nevill was so impressed by the buildings and the detailed knowledge of the guide that he had a rather extended discussion with him, and at the end he intended to purchase some information on Alsfeld and its history. But since we were in Germany on a Saturday afternoon, it was impossible to buy a book or even a picture postcard. All shops were closed. Nevill's reaction was typical, 'I certainly will forget the impressive town hall but I never will forget that it is impossible to buy a picture postcard at a tourist place on a Saturday afternoon.'

On Saturday evening he received further evidence for his observation that the Germans have a tendency to obey rules. We went out for dinner with colleagues to a restaurant; Nevill ordered trout and with it a red wine, which we knew he liked very much. The immediate reaction of one of the colleagues was, 'Nevill, you can't drink red wine with the fish.' But Nevill solved the problem in a sovereign way; he stretched his braces and said, 'My dear friend, I can.'

Peter Edwards, Friederich Hensel, Nevill Mott and Ursula Hensel at the Garden House Hotel, Cambridge (1996).

Another enduring memory comes from 1982 at a conference in Berlin on ionic liquids. Nevill gave the introductory address and I had the lecture just after him. Some of the participating theorists did not like my interpretation of our experimental observations, which quite normally for a physical chemist, I based on thermodynamic considerations. The discussion became a little confusing. Even today I am not sure whether I did not understand the very sophisticated arguments of the theorists or whether they had problems with the thermodynamics of a physical chemist. Nevill solved the problem. He stood up and declared, 'The argument of Professor Hensel is so simple that it must be true.'

I also have a very vivid memory of the second day of the conference when we absented ourselves from the afternoon session for a tour of the Berlin Wall, which he saw for the first time. He became so sad that he did not talk with me on the way back to our hotel. Still during dinner he continued to express his disgust about the existence of the wall.

I also recall very clearly the hours after this dinner. Nevill was preparing a lecture he wanted to present at the physics colloquium of the Freie Universität Berlin, two days later. He asked whether I would be willing to translate into German some new English words, concepts from the field of disordered materials, because he wanted to present the lecture in German. So we were sitting up to midnight in a hotel room and I had to inform this Nobel laureate on how Germans pronounce the word *lokalisiert*.

His last visit to Marburg was in autumn 1991 at age 86. He gave a lecture on spin polarons and the newly discovered oxide superconductors, and he still took great pleasure in exchanging ideas on this fascinating topic. I remember that we spent three mornings discussing studies in the metal–nonmetal transition range of fluid cesium from our laboratory. These studies revealed electron–electron correlation effects which he believed may aid the understanding of the new high-temperature superconductors. My young graduate students and postdocs were fascinated by seeing him walking around and shaking his head slowly from side to side while explaining patiently his seemingly uncomplicated models, and this at age 86.

After lunch in our house, the morning's exertions had left me ready to take a nap, but Nevill showed no sign of exhaustion. Instead, he expressed a strong interest in revisiting the countryside he had toured by bicycle back in 1929. He liked the forest with its nice pine trees and the hills he had cycled up. I vividly remember that Ursula and I brought him to the Waldeck fortress, from where one looks over a lake and a beautiful valley. During the tea break in the restaurant of the fortress, we had the chance to listen to an enthusiastic description of his bicycle tour through the Thüringer Wald. Despite his strict diet, Nevill was tempted to order a large piece of heavy German *Buttercreme* cake. He enjoyed it very much and remarked how the taste had not changed in sixty years.

On the way back to Marburg we stopped in Bad Wildungen, a small town with an old church which we wanted to visit. He was interested in the

The Reading Apostle by Konrad von Soest: it depicts the first pair of eyeglasses in German art.

bright colourfulness of the paintings of Konrad von Soest, who lived in the first half of the fifteenth century. According to Nevill, the most impressive of these paintings was *The Reading Apostle* (pictured here). It shows the oldest example of eyeglasses to be recorded in German painting, and this so fascinated Nevill that he bought a great many postcards to send at Christmas.

I am very happy that Ursula and I had a chance to see him in the summer of 1996, shortly before he died. The visit was most enjoyable because we found Nevill, now over 90, still in a mood to discuss science, social problems, the importance of politics for the future of the world and, last but not least, the interrelation of science and religion. So we had the impression nothing changed. He still continued to take pleasure in putting people on the way to success, as he always did. We enjoyed very much the meal in the Garden House Hotel in Cambridge. The accompanying photograph shows Nevill in the restaurant with Peter Edwards, Ursula and myself; he still enjoyed having friends around him.

Ursula and I cannot find the right words to describe the sadness we felt when we heard that he had died. We will miss him.

PROFESSOR DR F. HENSEL is professor of Physical Chemistry at the Philipps-University of Marburg. He obtained his diploma at the University of Göttingen in 1962 and his doctorate from the University of Karlsruhe in 1966. His research interests include electronic properties of small metal clusters and molecular clusters, fluid metals and semiconductors in the vapour–liquid critical region, nucleation phenomena, and properties of liquid alloys with a major emphasis on the metal–nonmetal transition.

A Man for All Seasons

Jenny Acrivos

My husband, Andreas, and I first met Professor Sir Nevill Mott and Lady Ruth Mott at Patras, Greece, during the summer of 1964. Andreas and Nevill had been invited by the government of Prime Minister George Papandreou to give advice on the founding of the new university near the city of Patras. They must have given good advice because today the University of Patras is thriving and one of the contributors to this volume, Professor E. Mytilineou, teaches there. We were thrown together for a week at the Cavouri Hotel by the sea near Athens. Ruth, a Latin scholar, had a wonderful time looking for ancient coins; all got a chance to do some sightseeing with friends. Most important of all, I got the courage to talk to Nevill about science and the metal to non-metal transition on which he was an authority. I was studying the chemistry of metals dissolved in ammonia.

Nevill and Ruth invited us to visit them in Cambridge, and we did so in 1968. We had dinner with the Master and his Lady at Gonville and Caius College, we met their daughter Mrs Alice Crampin and a very friendly group. I also visited the Cavendish at Free School Lane, where he showed me his parents' school picture. I was extremely touched and less tongue-tied when I noticed the admiration he had for his parents; it meant a lot to me that he admired his mother as a scientist. They also took us to the seaside cottage at Burnham Market, where we all had a wonderful time and saw beautiful flower exhibits at the churches. I went voluntarily, knowing I had allergic reactions, but am afraid that I sneezed all the way and made everyone uncomfortable. In California we had taken Nevill to the UC Santa Cruz campus, surrounded by beautiful tall prehistoric redwoods, and to the Monterey Bay Seventeen Mile Drive, where we enjoyed picking flowers – but then I had antihistamines! The Santa Cruz campus at that time was not attracting too many students and Nevill explained it very simply that students need to be cheek by jowl. The place was idyllic, but they needed a bustling town, and eventually success came when the city got what we called hippies in the late 1960s. I also brought Nevill to my laboratory at San José State University.

My first sabbatical leave in 1970 was spent at the Cavendish Laboratory in Free School Lane. I really enjoyed my early morning walks from Clare Hall past Saint Botolph's to hear lectures at 8 A.M. I also spent a lot of time at the library, where I used to see Professor Sir Brian Pippard (he was the master of Clare Hall, where we lived, and Andreas was a Fellow there). My research was considerably influenced by Dr Abe Yoffe, who asked me to intercalate metal ammonia (my expertise) into molydenum disulfide, a material his group were investigating at the time. I am by nature very bashful and when scientists accept me, I lose inhibitions. I made lasting friendships with Abe, whom I admire tremendously, with my coauthors, Dr John Wilson and Professor Yao W. Liang, and with Yao's wife, Chu, who was helping with crystal growth in the laboratory. I simply had a ball and Nevill got used to finding me standing on a stool doing glass-blowing on top of a table in the old laboratory when he came looking for Professor E. A. (Ted) Davis. They were writing their famous book at that time. My husband used to get mad at me for going to the laboratory at all hours but then he was only a theoretical fluid mechanicist working in the laboratory of Professor George Batchelor. The success of my work then was really due to the devotion of Arthur Stripe, a master in many ways. He taught me some new glass-grinding techniques to obtain vacuum seals, using optically flat surfaces in the old glass shed. I shared an office with three more scientists on the cold top floor. I also wrote a paper with Nevill and started a long correspondence with him on the subject of metal to non-metal transitions; I really got to understand his theory from my chemical point of view. I used to love going through the bridge (of sighs) between the old PCS and the Austin Wing of the Old Cavendish. I also remember fondly the garden at Nevill's home, where we took tea while discussing our work. Nevill had left the Mastership at Caius, so he and Ruth invited us for Sunday lunch also at their home with Professor and Mrs P. W. Anderson.

We visited Cambridge the next summer in 1971, just after a harrowing experience where our niece Maria had open-heart surgery at the age of 5.

Jenny Acrivos with Nevill Mott and Abe Yoffe (1984).

A contemplative Jenny Acrivos with Nevill (1994).

Coming to Cambridge and Clare Hall and Free School Lane was just like coming to a safe harbour. We both felt at home in Cambridge from then on.

When visiting Andreas' family in Greece we would always stop in Cambridge. In 1974 I was invited to talk at a meeting of Greeks from abroad (on account of my name only, since I was born in Cuba of Hispanic parents). The next sabbatical leave in 1977 found us again at Clare Hall and at the New Cavendish. Professor Tabor was the head of the PCS and Abe was training the future luminaries (for example, today's Cavendish Professor, Richard Friend) and had harboured Nevill in his laboratory in the Mott Building. Abe influenced my research again; he had saved my glassware and given it to Dr Alan Beal, who had just finished his PhD. Also Abe asked me to write a review article on intercalation chemistry and sent me to the University of Sussex, where Gordon Tatlock and I did the first work on the intercalation of hydrazine into tantalum disulfide under the electron microscope. It was great fun to do science in England again, especially because Nevill was on the same train to London, and then Brighton, to give a seminar at the university. I was carrying about a gram of ultrapure hydrazine under vacuum and one of Abe's precious crystals (I would not dare do it today).

The chemists were getting together in what we called Colloque Weyl V at Aviemore, Scotland in 1980. Nevill was invited to come to talk to us. He was in great form. Everyone was trying to get to understand antiferromagnetism and the new cage alkalide compounds that Professor Jim Dye (Michigan State) had prepared. After the meeting, Nevill rented a car and asked me to drive him to the Old Cairnhorn. It was quite cold and the mountains were extremely scary to drive through, but we made it.

The next vivid influence Nevill had on my life was at the meeting called by Uli Schindewolfe and Professor Fritz Hensel (Phillips University, Marburg), both chemists at Berlin in 1982. My heart was broken at the Berlin Wall, the inspectors had such looks of hatred as we crossed the border, but I shall never forget the smell of *Berlin Luft*; it was June and the trees were in blossom. Then I realized that it was time to get the physicists and the chemists together, doing work in the same area, so that we could start influencing each other directly. I proposed this to Nevill; he was receptive as usual, and the result was the NATO DAVY Advanced Science Institute. I wrote the proposal but asked both Abe and Nevill to cosponsor it. Yao Liang was the heart and soul of the meeting. We were housed at Gonville and Caius and had a marvellous time. It was my third sabbatical in Cambridge and I managed to bring together all those who had influenced me so much. Looking back it seemed such a natural thing to do. I recorded the lectures for posterity on tape. These tapes are now the property of Gonville and Caius, as I left them with Yao Liang when I came back to Cambridge for Nevill's memorial service on 16 November 1996. The photograph on p. 245 shows the editors of the DAVY NATO ASI: Abe Yoffe, Nevill and myself.

The late eighties signalled the appearance of superconductivity in cuprates. We had worked with intercalated tantalum disulfide (a mere 5 K superconductor) in the late seventies and also with a layered organic conductor which showed a magnetic transition near 7 K. A colleague in this work was Stuart Parkin, whom I had met at the Cavendish as a beginning graduate student with Richard Friend. He came to work with me in 1980 and obtained the first X-ray absorption spectra of layer intercalates, stayed longer than expected (two weeks at my house after the experiments were over) and we wrote the work up next summer, when I came back to Cambridge. Yao Liang provided an office for Nevill in the IRC for Superconductivity at the Cavendish after Abe's retirement. Nevill paid his rent by becoming interested and very productive in unravelling the mechanism for superconductivity. Together with Professor Alexander Alexandrov, he proposed the bipolaronic mechanism for superconductivity and asked me if I knew of any superconductors which showed the presence of triplet states. I went back to my organic metal. By that time I knew more about detecting the transition to superconductivity in cuprates using energy loss in zero field, so I verified the relevant transition near 10 K and was able to associate it with the triplet state. I dedicated this work to Nevill for his ninetieth birthday.

The mentor and a disciple on their last meeting (p. 246) have their hands clasped similarly in thought, pleading for understanding. I continue to work and dedicate my work to the person who has influenced my life almost as much as my mother, Lilia, and my husband, Andreas. I sincerely hope and pray to my creator to allow me to work for as long as Nevill did.

PROFESSOR J. V. ACRIVOS, a solid-state chemist, was born in La Habana, Cuba. She obtained a DSc from the University of Habana and a PhD from the University of Minnesota. She has been a Visiting Fellow Commoner at Lucy Cavendish College and Trinity College, Cambridge, and NSF Visiting Woman Professor at the University of California in Berkeley. She is currently Professor of Chemistry at San José State University, California.

Encountering Sir Nevill

John Goodenough

In the spring of 1946 I was called home from my posting in the Azores as a meteorologist in the US Army Air Force to be given a fateful surprise, the choice to study either physics or mathematics at the University of Chicago. I had been toying with the idea of returning to study international law; the need for a world order under the rule of law appeared so compelling to those of us whose lives had been interrupted by the Second World War. But in Chicago I was rescued from naivete of the political realities of our day by a flashback to an evening in 1941 when, while reading Alfred North Whitehead's *Science and the Modern World*, I had determined I would read physics should I have the good fortune to return from the war and have the opportunity to go back to school. Science was destined to play a critical role in the intellectual dialog of my generation, and physics seemed to be the logical entry into that dialog. Now I stood before that door, and I accepted the challenge.

On registration the next day, I was asked by Professor Simpson, 'Why does one of your age want to begin the study of physics? Don't you know that anyone who has done anything significant in physics has already done it by the time he was your age?' Clearly I was not an obvious candidate for a career as a physicist's physicist, but I persevered with a modest thesis in metal physics under Clarence Zener.

It was while a student at Chicago that I first encountered Nevill Mott, as the coauthor of essential texts and then, in about 1950, as a seminar speaker discussing why NiO was an antiferromagnetic insulator rather than a metal. In his characteristic style, he had put his finger squarely on a fundamental problem in solid-state physics and drawn it to our attention. I was to revisit that problem some years later.

The impact of Mott on my development as a student became evident on the presentation of a principal result of my PhD thesis at an American Physical Society Meeting. I had written my thesis away from the University of Chicago while working at the Westinghouse Research Laboratory in East Pittsburgh, and key examiners-to-be from Chicago came to hear how I would do before a critical audience. I discussed how the interaction of the Fermi surface with the Brillouin zone boundary influences the evolution with electron/atom ratio of the axial ratio c/a of the hexagonal Cu–Zn system. At the end an old gentleman – he turned out to be Brillouin himself – stood up in the first row and said, 'That's all very well, young man, but the close-packed hexagonal structure is not a Bravais lattice, and you have not presented the correct Brillouin zone!' I had used the first Brillouin zone presented by Mott and Jones, not that given by Brillouin and reproduced in the text by Fred Seitz. Devastated, I returned to Pittsburgh to rescue my thesis that weekend by proving Brillouin mathematically correct, but physically wrong. Mott and Jones had correctly identified the first physically meaningful Brillouin zone of the hexagonal close-packed lattice.

Upon graduation, I went to the MIT Lincoln Laboratory to work on the development of a ferrospinel with a square B–H hysteresis loop for the coincident-current magnetic-core memory of the digital computer. I was strongly influenced at that time by Pauling's *The Nature of the Chemical Bond*, Rick Bozorth's *Ferromagnetism* and the papers of Louis Néel on anti-ferromagnetism and ferrimagnetism. The work of Van Vleck and his student Phil Anderson, with whom Sir Nevill Mott shared the Nobel prize, introduced important mathematical components, as did the landmark papers of John Hubbard that clarified the Mott–Hubbard picture of NiO, just as I was exploring the transition from localized to itinerant electronic behavior in the transition metal oxides. It was my work on the narrowband problem and on the lattice instabilities encountered at the transition that induced Nevill Mott to initiate a correspondence with me.

I have forgotten what the problem was that Nevill posed for me in that first correspondence. I only remember my astonishment that he would solicit my opinion and that I would struggle for a week to come up with a reasonable response; he would have another question in the return post! Once the problem was resolved to his satisfaction, he offered me a coauthorship. As it was not my problem, and my contribution had been minimal, I declined. However, I fear he may have been offended that his generosity was not more graciously received.

At about the same time, Nevill Mott visited me at the Lincoln Laboratory. When I introduced a problem on which we were working, he did not wish to hear our solution until he had first become fully engaged with a resolution of his own. With this approach, he enjoyed himself while acting as a useful critic.

On one occasion, we had the pleasure of entertaining Nevill Mott for a weekend in our home in New Hampshire while he was on his way to a Gordon Conference. He proved a delightful and gracious house guest. My wife, Irene, was teaching history and philosophy at a community college in Boston, and she explored with him the art of communicating to students the great concepts of our Western civilization. Mott spoke enthusiastically of the experiment in adult education in the United Kingdom that had recently been launched as the Open University. My wife was later to teach in that institution when I was Head of the Inorganic Chemistry Laboratory at Oxford. Although we shared with Mott an interest in religion, we never succeeded in engaging him in a sustained dialog on that subject. He gave us the impression that strongly held opinions on theology were more formative for him than personal exploration of the moral domain.

My one joint publication with Sir Nevill came about as a result of a simultaneous visit to Bordeaux. At my instigation, the group of Demazeau had prepared $LaCuO_3$ under high pressure, and his colleagues in Bordeaux had measured the temperature dependence of its paramagnetic susceptibility for comparison with that of $LaNiO_3$. Mott was immediately engaged with the data, which illustrated a distinction between a mass enhancement of the susceptibility in metallic $LaCuO_3$ and a Stoner enhancement in metallic $LaNiO_3$. The chemists felt more comfortable asking me rather than Sir Nevill to clarify the distinction, but I don't know to what extent I succeeded. An engagement with the internal world of Sir Nevill Mott was always an intellectual challenge and an enlightening experience.

Abe Yoffe was a wonderful experimental colleague for Sir Nevill Mott; and at the warm celebrations he organized for Mott's seventieth and eightieth birthdays at Cambridge, it was evident that Mott had touched many in a personal way, just as he had influenced me. He was a man totally engaged by the challenge of solid-state physics, and he kept actively probing the core issues of the subject with excellent collaborators to the end of his long life. He was one of the pioneers of a generation that introduced physics into the field of metals to transform a traditional empiricism into a modern science. Those of us who have followed owe him a great debt.

PROFESSOR J. B. GOODENOUGH was born a derivative US citizen in Jena, Germany, in 1922. He studied classics at Groton School and majored in mathematics and philosophy at Yale University before entering military service in 1942. After the war, he obtained a PhD in physics at the University of Chicago and was a Group Leader at the MIT Lincoln Laboratory from 1952 to 1976, Professor and Head of the Inorganic Chemistry Laboratory, University of Oxford, from 1976 to 1986, and has been a Professor of Engineering at the University of Texas at Austin since 1986. His research contributions have been primarily concerned with the electronic and magnetic properties of transition metal oxides.

The Enquiring Chemist

Peter Edwards

I first met Sir Nevill Mott in late 1977, a very short time after the public announcement of his Nobel prize in physics. This was already six years into his supposed retirement, a state of mind and a physical condition to which we all know he never adhered, right up to his 91 years. However, I had known the name Mott for some considerable time, not our own NFM but his father, C. F. Mott, late Director of Education in Liverpool.

As a teenager in Liverpool I regularly passed the C. F. Mott College en route to my own school. Of course, I could never have imagined that C. F. Mott's offspring would have such a profound influence on my own future scientific life and career. My interaction with Mott junior – admittedly then at arms length only – began during my PhD at Salford University.

A chemist by training – but increasingly a physicist (some might suggest a token physicist) by inclination – I became increasingly aware during the early 1970s of Mott's scientific writings. During the latter period of my PhD (around 1974), at the instigation of my supervisor, Ron Catterall, I summoned up the courage to write to Nevill at Cambridge, asking him for any of his reprints that he might think appropriate to my own thesis project on rapidly quenched metal–ammonia solutions.

The request from a mere graduate student at Salford was, to my delight, met instantly by the arrival, from Cambridge, of a large collection of scientific articles dating back to Nevill's early work in 1949 (incidentally, the year I was born) on the metal–nonmetal transition. I vividly remember my many attempts in 1974 and 1975 to peruse – not yet even to digest or understand – this vast and erudite scholarship. For here I found Nevill's remarkable writings on the transition to the metallic state, the Mott criterion, variable-range hopping, the minimum metallic conductivity and many more, inspiration and ideas which still remain central to my own thoughts to this very day. It is true to say that I was overwhelmed by the sheer breadth, and depth, of the physics and chemistry contained in those beautiful pages. This multitude of papers accompanied me to Cornell, where I was to spend two years (1975–77) with Michell (Mike) J. Sienko in the Baker Laboratory of Chemistry.

My sheer frustration with my own shortcomings in trying to come to grips with Nevill's writings acted as the catalyst for me to attend every possible solid-state physics and chemistry course at Cornell. In physics I would single out the beautiful lectures given by Neil W. Ashcroft, which formed the basis of his classic text with N. D. Mermin.[1] In chemistry, the beautifully transparent lectures from Roald Hoffmann, Mike Sienko and Ben Widom also remain a vivid memory of a most stimulating period at Cornell.

My first 'formal' encounter with Nevill was on Tuesday 22 November 1977 in the Inorganic Chemistry Laboratory (ICL) on South Parks Road, Oxford. Upon my arrival at Oxford in September of that year, I had received a note from Nevill indicating that 'he knew of my work with Catterall and Sienko' and would like to come and discuss science. There I was, a young postdoc receiving the new Nobel laureate in physics, and entertaining him in the communal postdoc office in the ICL, known affectionately as the rat hole. In the two weeks before Nevill's visit, I spent every possible hour perusing – in hindsight, though, I must say revising – every aspect of solid-state physics from Ashcroft and Mermin which I thought might come up in the forthcoming Mott 'exam'. Nevill's first enquiry, however, left me temporarily reeling, 'Edwards, tell me about the chemistry and oxidation states of tungsten and bismuth.' Ashcroft and Mermin was rapidly put to one side; the chemistry of the elements was in the ascendancy!

I also recall our second meeting, some weeks later. Following our further discussions at the ICL, Nevill was to present a special lecture to the under-graduate Physics Society; I believe this was in the Clarendon Laboratory.

The young undergraduate who introduced Nevill was certainly not over-awed by his eminence, nor did he stand on ceremony. After an unholy (short) introduction, 'Today, we welcome Professor Mott from Cambridge who has just won the Nobel prize,' he rapidly moved on to remind, one might almost say reprimand, the audience that 'all chairs have to be stacked away imme-diately after the lecture has finished.' Nevill, unruffled, rose majestically to his feet, handkerchief in hand, 'Thank you, my dear boy, I never did like long introductions.'

The excellent two-week NATO School organized by David M. Find-layson and David P. Tunstall and colleagues at St Andrews during August 1978 was the first time that I was able to observe Nevill, as one might say, at close quarters in a scientific conference. Two memories spring to mind. The first was his undoubted taste for Glenfiddich and other fine malts. I still recall parading around St Andrews in heavy summer rain with Nevill and his huge, brightly coloured golf umbrella (I think I counted eight colours) presented to him by David Findlayson. The second overriding memory of St Andrews relates to science. In response to a particularly complex ninety-minute theory discourse, I recall Nevill rising to his feet, perennial brown case open on desk, handkerchief brandished, and with that stilt-like manner and beautiful angled delivery, noting to the lecturer, 'My dear friend, before your lecture I was unsure of the magnitude of this fascinating effect, but now I have to say I'm unsure even of its sign!'

My own transition (1979) from postdoc at Oxford to demonstrator at Cambridge was a particularly pleasant experience, made even more straight-forward by Nevill's response to my enquiry about Jesus, my prospective college. He responded that this was a fine College, 'Of course, you have Alan

Cottrell and John Adkins ... and the parking is excellent.' This combination of observations sealed my decision!

Another enduring memory of my period at Jesus was the 1983 visit by C. N. R. (Ram) Rao, first Jawaharlal Nehru Visiting Professor at Cambridge. He was accompanied by Nevill and the visit was hosted by Sir Alan Cottrell and myself. After a sumptuous meal and an endearing after-dinner combination with the Fellows, where the wine, port, madeira and whisky flowed until well past eleven o'clock, Ram Rao and I wondered whether we should escort Nevill back to Gonville and Caius College. 'Thank you, but no, ... I will be driving back to Milton Keynes tonight,' replied Nevill. Ram and I exchanged an incredulous glance, and walked the 78-year-old driver back to his car. That was also the period when, confronted by his ageing car, peppered with an advanced state of rust, Nevill pointed to the percolating clusters of orange-brown iron oxides and noted to me, 'There you have a Mott transition!'

How we already miss him, not only with a deep sadness, but also with a chuckle and a grin. How often do we all recall opening letters from Nevill obviously destined (and addressed) to an institution such as Bell Labs, only to note that 'Dear Gordon'* had been crossed out to read 'Dear Peter'! With anyone else one may have been offended, but with Nevill this event brought a warm smile – and indeed an inner satisfaction – knowing that one was part of his orchestrated, handwritten 'World-Wide Web' of cross-linked friends and collaborators, singled out for 'private circulation to solicit comment'.

We surely all remember those instances with Nevill when somehow one just sensed that his attention was perhaps wandering. I vividly recall meeting with him in Cambridge in 1990 to discuss high-temperature superconductivity. Nevill courteously asked me to summarise my own views of the present situation in terms of the chemistry and physics of the problem. Deep into my impromptu tutorial, I noticed his eyes flashing rapidly around the room. Taking advantage of my staged prompt to our guru for an incisive, penetrating comment, Nevill, handkerchief poised at his lips, responded, 'Peter, you will join me for lunch, won't you?'

In the last years of his life, the enormous intellectual challenge of high-temperature superconductivity was indeed central to Nevill's science. This period marked a particularly pleasant and stimulating interaction with him. It was also characterised by a considerable body of correspondence between us, and very lively and thought-provoking discussions in Cambridge with A. Sasha Alexandrov (now Loughborough).[2] Sadly, we were only part-way through our joint paper 'Are High Temperature Superconductors Metals?' in the summer of 1996. This very question reminds me that Nevill always used to tease me that my own sporadic forays into theory inevitably resulted in many questions, but not always answers!

A particular favourite of mine has been the venerable enquiry, "What is a Metal?" Mike Sienko and I posed this question in 1983. Recently I have

* Dr G. A. Thomas, Bell Laboratories, Lucent Technologies, Murray Hill, New Jersey.

Th

Dear Peter

I've thought a lot about "What is a metal" + I think one can only answer the question at $T = 0$. Then a metal conducts, + a non-metal doesn't. This includes heavily doped semiconductors as metals, of course

I still don't know how to calculate the transition for $La_{2-2}Sr_x CuO_4$!

Yours
Nevill

Letter to Peter Edwards, 9 May 1996: Nevill answers the question, What is a metal?

become aware that the very same enquiry was posed much earlier by Sir Alan Cottrell (then Dr Cottrell) in 1946 during a lecture to the Birmingham Metallurgical Society! Nevill's very last letter to me (reproduced here) on 9 May 1996 will always bring a wry smile to my face; here he was, at age 90, now acquiescing to this very same perplexing question so often posed by the chemist Edwards and noting, 'I've thought a lot about "What is a Metal".'

My very last social meeting with Nevill was on 5 July 1996 in the Garden House Hotel, Cambridge, with our close mutual friends Ursula and Friederich Hensel (pictured on page 240). I am sure that both Ursula and Friederich will vouch for the fact that, even into his nineties, Nevill still revelled in the company of friends.

I will always feel immensely proud, and privileged, to have been part of Nevill's network of friends and colleagues, worldwide, with whom he shared his kindness and generosity, his towering scientific insights, his mischievous humour and his immense love of life. His greatness touched us all.

I have heard many people speak of Nevill's passing as the end of an era. Of course, this has to be correct, but we should see this as the celebration of an era – the Mott era. For this, we thank you Nevill; we miss you already.

Acknowledgments

I would like to thank David E. Logan and Ray L. Johnston for their help in producing this appreciation.

References

1. N. W. Ashcroft and N. D. Mermin (1976) *Solid State Physics*. Philadelphia PA: Saunders.
2. P. P. Edwards, N. F. Mott and A. S. Alexandrov (1997) The insulator-superconductor transformation in cuprates. *Journal of Superconductivity*, in press.

PROFESSOR P. P. EDWARDS, FRS, is Professor of Inorganic Chemistry at the University of Birmingham. Following undergraduate and postgraduate studies at Salford University, he held appointments at Cornell, Oxford and Cambridge, before moving to his present position in 1991. He has made various contributions to condensed matter science, with a common theme being the metal–nonmetal transition.

Sharing a Whisky

John Meurig Thomas

Intellectually, my first meeting with Nevill Mott occurred in the mid 1960s when one of my bright research students at the University College of North Wales, Bangor, where I was then a lecturer in chemistry, persuaded me that space-charge-limited currents might be a useful method of probing electron and hole traps in organic molecular crystals, such as anthracene. This led me to non-ohmic currents and to the existence of Child's law, which were until then quite unknown to me. By chance, I stumbled upon the beautiful monograph *Electronic Processes in Ionic Crystals* by N. F. Mott and R. W. Gurney, and shortly thereafter I felt infinitely better versed in solid-state phenomena than I had been hitherto.

It was not merely the elegance and economy of the 'proofs' in that text that impressed and exhilarated me – Child's law was derived in two lines – it was also the wealth of discriminating diagrams and figures culled from original scientific papers that also left a profound impression. I vividly recall discovering in that text the pioneering work on soft X-ray spectroscopy and the direct determination of electronic band structures of metals and other solids reported (not long before the text of Mott and Gurney went to press) by H. B. W. Skinner. It was as a result of reading *Electronic Processes in Ionic Crystals* and Skinner's work in *Phil. Trans.* that I knew precisely how to set about determining the band structure of graphite, diamond and a number of other solids once an X-ray induced photoelectron spectrometer as a new tool became available to me in 1970.

Mott's interpretation of the observed rates of the oxidation of metals (published with Cabrera) along with his explanations of the photographic process and of the spectra of colour centres soon became part of my undergraduate lecture courses. And his charming booklet on metals and dislocations left its mark upon me.

But apart from a very brief encounter in Marshall Stoneham's office at Harwell in the mid 1970s, I did not really get to know him until my family and I moved from Aberystwyth to Cambridge in 1978. One autumnal evening, as I was busy picking apples from trees at the foot of our long garden in Sedley Taylor Road, I sensed that someone had approached me quietly from behind. When I turned round, I saw it was Mott, exuding friendliness and charm. 'I gather that you are our new Professor of Physical Chemistry, Linnett's successor. Tell me what kind of science you do.' After some ten minutes of rather confusing explanation on my part – I was rather nervous because I was conscious of speaking to a great man of exceptional perspicuity and charm, – he beamed and said, 'We have some things in common.' After a brief discussion about the beauty of, and the stimulus provided by Cambridge, he started to walk homeward, 'Come and knock at my door if you would like to share a whisky with me.'

Thereafter our friendship grew. On more than one occasion when I requested that he might give seminars to the physical chemists in Lensfield Road, he instantly obliged. Likewise he responded enthusiastically to requests for synoptic reviews for journals with which, as coeditor, I was associated. (The promptness with which his promised texts arrived contrasted sharply with the dilatoriness of many a younger contributor.)

In the last ten years of his life, our discussions tended to centre on the mechanisms of superconductivity in 'ceramic' oxides, a subject to which I could contribute very little. But I always relished reading what insights Mott provided to the phenomenology of these intriguing solids. I remember one of his reviews on this subject, published in the early 1990s, starting with the lovely assertion, 'As long ago as 1988 it had become clear ....' No one who knew Nevill Mott would suspect any hint of Stephen Potter one-upmanship in this opening. Here was the man who had raced ahead of his contemporaries sixty years earlier – by, as he put it, simply applying the Schrödinger equation to an obvious problem – still indulging in genuine cerebral curiosity.

In the last five years or so I received from him one or two articles on religion, written with the same absolute honesty that imbued his scientific endeavours. One argument of his readily springs to mind. 'If you ask me why I believe in God, I suppose the true answer has to be: Because I want to.'

Long before I ever met Nevill Mott, friends and acquaintances had told me how charming he was. 'In my early days as a research student in Bristol Mott's messianic glow was apparent to us all,' thus spoke Trevor Evans, decades later. When I finally got closer to Nevill Mott, it was the combination of his stunning ability, English charm, endearing absent-mindedness, insatiable curiosity, human decency and towering intellectual authority that impressed me. She was unique, and he is unforgettable.

SIR JOHN MEURIG THOMAS, FRS, is Master of Peterhouse and Professor of Chemistry at the Royal Institution, of which he was Director from 1986 to 1991. In 1978 he moved from Aberystwyth, where he was Head of the Department of Chemistry, to head the Department of Physical Chemistry in Cambridge.

A Man and a Scientist

Marie-Luce Thèye

I have known Professor Mott for about a quarter of a century. I never was one of his students and I never worked with him or even on subjects he was working on himself; we only met rather briefly two or three times a year, in various places over the world but mostly in Cambridge or Paris, and yet I miss him terribly, both as a scientist and as a man. Probably like most of the people who, for one reason or another, happened to meet him and became somewhat close to him, what I felt towards him, since the very beginning, was a mixture of deep admiration and sincere affection. He represented a kind of landmark in my life. I miss his enthusiasm, his insatiable curiosity, his sense of humour and his cheerful 'Hello, Marie-Luce!' with which he used to greet me.

I first met him at the International Conference on Amorphous and Liquid Semiconductors which he organised in Cambridge during September 1969, and which initiated a series of successful conferences that continue to cover this field. What happened then perfectly illustrates why I was right away *sous le charme*. After obtaining my *doctorat d'état* in Paris in 1968, I was searching for a new field of research. Professor Jan Tauc, who had already started to study amorphous semiconductor thin films in Czechoslovakia in collaboration with other colleagues in Eastern Europe, drew my attention towards the problems specific to this new class of disordered materials during his visit to Paris as an invited professor at the beginning of 1969. These problems immediately attracted me by their novelty and excited me by their complexity, and I decided to attend the Cambridge conference in order to obtain more information. However, for various reasons my application arrived at the very last moment, and to my bitter disappointment, I received a negative answer justified by the limited accommodation facilities offered by the colleges of Cambridge. A few days later, however, somebody in the laboratory called me, saying, 'A certain Professor Mott is on the phone, asking for you.' I just could not believe it, but there he was *en personne* and, very nicely, he told me that he had succeeded in finding a college room for me for the first three nights of the conference, and that I could come, if I still wanted to. I probably stammered a few words of thanks in my very poor English, and I hung up the phone, walking on air.

I must say at this point that Mott's name was far from being unfamiliar to me. My thesis work was dedicated to the optical and transport properties of noble metals and their alloys, and the Mott and Jones book had almost become an object of veneration for my family, because I carried it with me

259

Amorphous brethren at a conference in Prague (1987): Marie-Luce Thèye stands in the centre.

everywhere during all my vacations, and I must confess, an object of jokes (and of shame for me) because I very rarely opened it (during the vacations, that is). Many years later, as he was recalling his official activities after the Second World War, Professor Mott told me that Professor R. V. Jones, another member of the English delegation visiting German laboratories, was a little irritated because too many German solid-state physicists persisted in confusing him with the Jones of the all too famous *Metals and Alloys*. It is certainly a common experience for many young scientists to discover that the authors of venerated books are real persons whom one can meet and talk with if one is lucky enough. Hence the emotion caused by this telephone call about a rather trivial problem. Of course, I rushed to Cambridge, and on the first day of the conference, I asked one of my English friends, Professor John Hodgson, to introduce me to Professor Mott, because I wanted to thank him for his kindness and personal involvement. After the introduction, Professor Mott said, 'I am very happy that you could come. Don't worry, I will find another college room for you for the last part of the conference. But in all events, I can always ask X (somebody of his family) to accommodate you in her house for a few nights.' I was so astounded by this totally unexpected solution to my 'problem' that unfortunately I do not remember who was the very generous person he was referring to. But I am still deeply grateful to that person for her 'virtual' hospitality, which eventually she had no opportunity to exercise since I was indeed able to move to another college.

The anecdote does not make the whole story. I had already attended several conferences and workshops. But this one in Cambridge was for me a real eye-opener. There were rather few participants, but all of them were so enthusiastic, there were so many ideas crackling around, there was such a liberty of discussion, that I was totally fascinated. I will never forget Professor Mott challenging everybody with his loud voice and communicative excitement, putting up new arguments on any subject, just for the pleasure of it, stepping over a bench or a table if necessary to come close to his partner and try to be more convincing. For the first time, I realised what enjoyment doing science could be!

Professor Mott was present when I reported my first results on amorphous germanium at a small conference organised in Cannes by Benoy Chakraverty. He recognised me but we talked only briefly at that time. So, what a surprise for me when, a few weeks later, he knocked at my door at the Laboratoire d'Optique des Solides, without notice, and said with amazing simplicity and total absence of pretension or affectation, 'I was interested by what you said in Cannes. But, you know, I am not so familiar with optical properties. Would you then be kind enough to explain to me again the results which you obtained, and also what you expect from such studies?' So I did, with an immense pleasure and pride; he looked so happy to learn something new! What happened to me on this famous day is very representative of the way he behaved with anybody in our community who had attracted his interest by an idea out of the common understanding, by puzzling experimental results or by an unconventional interpretation of some data. Many times have I seen him 'interview' some young, or not so young, scientist, no matter their age, sex, nationality, etc., no matter how famous or how unknown! He could also completely ignore what the same people were doing for months, because it was far from what he had in mind at that time. And then, one day, he would come back to them saying, 'Could you explain your idea to me?'

In 1986 we made a journey from Paris to Nancy. He had been asked to sit on the thesis jury of a young woman from Algeria, Nabila Maloufi, who had undertaken at the Laboratoire de Physique des Solides of the Nancy I University very nice experimental work on the low-temperature conductivity of amorphous silicon–tin alloys, and had analysed her data with the different available models, including the famous $T^{-\frac{1}{4}}$ law. He agreed with his usual enthusiasm, not only to attend the thesis presentation, but also to write a report on the manuscript. I was the second referee, and we found it convenient to travel together by train. Once again, I was amazed by the energy, the curiosity, the cheerfulness and the kindness he showed during the long trip (he talked almost all the time), the tour of the town and museum, the various ceremonies he had to endure from Nancy's authorities as a visiting celebrity (he was 81 years old then). But what I really want to recall about this memorable journey is a remark which he made on our way back to Paris. He had been impressed by the quality of both the work and the presentation of Nabila, and he talked to her with obvious pleasure during the lunch organised afterwards in a small restaurant close to the university, while enjoying the food and wine *en connaisseur.* In the train back to Paris, he started explaining to me how much he had lately been concerned by the problem of religion, because he felt that his life in science had led him to neglect all forms of spirituality, and why he had chosen, or was about to choose, the Anglican Church (which allowed him to preserve enough liberty of thinking). And suddenly, after a long silence, he said, 'I was remembering what this bright young woman told me at lunch, that probably she will not be able to do research anymore when returning to her country. You see, Marie-Luce, I have tried to imagine what hell could be; and one of the possibilities I found is not to do science anymore.'

I saw him again in 1993 at the International Conference on Amorphous Semiconductors. He was so cheerful, so pleased with the food and wine, so eager to participate in all the conversations, to laugh at all the jokes, so happy to be once again among 'his' people, and he himself held the *place d'honneur* at the conference banquet; the event had returned to Cambridge after twenty-four years.

DR MARIE-LUCE THÈYE is Directeur de Recherche at the French Centre National de la Recherche Scientifique (CNRS), and is presently the Director of the Laboratoire d'Optique des Solides of the Université P. et M. Curie, Paris 6, Unite Associée on CNRS 781. She has been working on the relationships between the optical properties and the microstructure of amorphous semiconductors (amorphous silicon, germanium and III–V compounds, amorphous chalcogenides, hydrogenated amorphous carbon) since the early seventies.

Non-Crystalline Solids: order in disorder

Walter Spear

Nevill Mott's wide-ranging contributions to the field of non-crystalline solids were undoubtedly among his major scientific achievements. During the 1960s and the 1970s his work led to new ideas and concepts on which much of our present understanding of electronic processes in these materials is based. Throughout this period he maintained close personal contact with experimentalists in the field by extensive correspondence and frequent visits to laboratories in Britain, Europe and the United States. He became a central figure in the international research effort and stimulated the rapid growth of the subject through direct communication and discussion of the latest results, through his papers and through the well-known book he wrote jointly with Ted Davis.

I first met Nevill in 1960, when he visited the Physics Department of the University of Leicester, where I had started experimental work on electronic transport in low-mobility solids, both crystalline and non-crystalline. Nevill was fascinated when he saw on the screen of our ancient oscilloscope the transit of excess electrons and holes across a thin evaporated film of amorphous selenium, a material which became of importance in first-generation photocopiers. Although the results of these time-resolved transport experiments were clear-cut, their interpretation in terms of basic electronic interactions presented many challenging problems at the time. The Leicester visit marked the beginning of an extensive correspondence between Nevill and me which, depending on the research results, continued well into the 1980s. Nevill had a remarkable ability to absorb essential aspects of new results; shortly after a visit or discussion, a series of handwritten letters would generally arrive, some quite brief, perhaps with a new idea or suggestion, others extending over several pages with detailed attempts at interpretation.

During the 1960s there developed increasingly close contact between Leicester and Cambridge. Some of my work was then concerned with transport in molecular crystals. Nevill was particularly interested in the experimental evidence for small-polaron formation in orthorhombic sulphur crystals and also in the work on electron transport in solid and liquid argon, relevant to the effect of disorder at a mobility edge.

My appointment in 1968 to the Harris Chair of Physics at the University of Dundee gave me a timely opportunity to set up a new research laboratory, specifically for the experimental study of amorphous solids. In this venture my close collaborator, the late Peter Le Comber, and I were strongly motivated by Nevill's work and realised the need for decisive experimental tests of new

Nevill relaxing in the garden of Walter and Hilda Spear's cottage near Dunkeld (circa 1974).

concepts in the field. During the 1970s our laboratory became a centre for research on amorphous silicon, a material of considerable fundamental interests which later found many applications. Nevill was a frequent visitor to the laboratory and to my home in Dundee, and he rarely left us without proving himself to be the genuine absent-minded professor, leaving behind a shirt or a handkerchief. His interest and encouragement were a great help to the development of our research group.

In Summer 1972 I organised the Thirteenth Scottish Summer School in Physics at Aberdeen; its theme was electronic and structural properties of amorphous semiconductors. The three-week gathering attracted many younger workers in solid-state science. Nevill, one of the principal lecturers, thoroughly enjoyed the informal atmosphere. Each day before dinner the lecturers retired into a small staff room, well-stocked with a full range of Scottish malt whiskies for much-needed refreshment. Nevill developed a particular liking for Glenlivet and jocularly blamed me for may years afterwards for having 'ruined' him. Needless to say, a bottle of that magnificent spirit became a frequent birthday gift.

With the growing interest in amorphous solids, Dundee and Edinburgh Universities started a joint one-year postgraduate course on electronic properties and applications. In each session we invited Nevill to talk to the students about his latest ideas and results. The highlights of his visits, however, were the informal discussions, sitting around the blackboard in my study. PhD and MSc students in the group were all keen to talk to Nevill about their research projects and the problems involved. He always treated them with courtesy and understanding and never resented it when he was occasionally proved wrong in the discussion.

One year I persuaded him to give an informal talk to the faculty on his life in science and on some of the famous physicists he had met. It was a great success and afterwards at tea he was joined by a very attractive female honours student, who delighted him by remarking, 'Professor Mott, that was a fascinating talk – I could listen to you for hours!'

In addition to correspondence and laboratory visits, national and international conferences were for him an important means of communication. Nevill was one of the founders of the informal Chelsea meetings where, just before Christmas, colleagues and research students discussed their latest results on non-crystalline solids and liquids. These meetings greatly stimulated progress and soon attracted colleagues from France and Germany.

I remember a conference at Clausthal, a small university town in the Harz Mountains. We were treated to a simple but excellent conference dinner of venison, vegetables and an unlimited supply of beer. Nevill appeared greatly to enjoy himself and, as he was somewhat unsteady on his feet, my wife and I helped him back to his hotel. The next morning I complimented him on his capacity for German beer and he answered, 'You'd be surprised, Walter, how much beer I can drink – particularly if it's free!'

A somewhat embarrassing episode occurred during a conference on amorphous solids held in 1987 at Lake Balaton in Hungary, to which we had been invited. The Russian delegation was led by the late academician B. T. Kolomiets from the Ioffe Physical-Technical Institute in what was then Leningrad; over many years he had done very distinguished work on chalcogenide glasses. The conference outing was to a small town on Lake Balaton, well known for its International Memorial Park in honour of Nobel prizewinners. Nevill was asked to plant a tree next to a plaque inscribed with his name and achievements.

A large crowd of local officials, conference participants and others were assembled for the planting ceremony. Nevill stood in the centre, holding the small tree in one hand and a spade in the other. At this moment, Kolomiets pushed to the front, took hold of the stem and announced something to the effect that as a Russian colleague and friend, he insisted on helping Professor Mott. After an initial reaction of astonishment and annoyance, Nevill put on a broad smile and welcomed Kolomiets's help. This thoughtful gesture saved the situation!

Nevill was deeply interested in theology and religious thought; it is perhaps less well known that this interest extended to a lively appreciation and knowledge of ecclesiastical art and architecture. During one of the Rome conferences in the mid-seventies, Nevill and I took a day off for a carefully planned tour of the old basilicas in the city. We also shared an enthusiasm for the superb sixth-century mosaics in Ravenna and during my last visit there I managed to find him a good reproduction of the *Procession of the Magi* (from Saint Apollinare Nuovo), which gave him much pleasure.

Finally, a bon mot attributed to Nevill in the Bristol days. A talkative and somewhat tedious visitor at long last took his leave. 'Well,' said Nevill, 'it was nice to see you. Perhaps you will come back one day when you have less time.'

WALTER SPEAR, FRS, FRSE, was Reader in Physics at the University of Leicester until 1968 and then Harris Professor of Physics at the University of Dundee until his retirement in 1990. His main research interests lay in the field of electronic transport and optical properties of low-mobility solids, both crystalline and amorphous. He now devotes his time to reading, music and DIY problems.

Nobel Prize News in Marburg

Josef Stuke

It was a sad day for me when I read in a German newspaper that Nevill Mott had died on 10 August in a London hospital. Since the brief article didn't give any information on the cause of death, I called Walter Spear in the afternoon. He told me that Nevill had had a heart attack and I was relieved to hear that he was spared a longer suffering.

Quite vividly I remember how impressed I was by Nevill's personality when I met him for the first time in 1966 at the conference on low mobility solids in Sheffield. I was still at the Technische Hochschule in Karlsruhe and we were working there on the change of electrical and properties at the transition from the crystalline to the amorphous state. I presented at the conference a review of these investigations; after my talk, Nevill expressed his great interest in the results and we had a long discussion on the interpretation. That was the beginning of a close scientific and personal relationship which lasted thirty years.

A couple of months later Nevill came to Karlsruhe and we continued our discussions there; after my call to the University of Marburg in January 1967, many meetings there and in Cambridge followed. I remember quite well my first visit at the old Cavendish when he was still Cavendish Professor, head of this famous laboratory. He took me around the old building and it was an unforgettable experience to see Maxwell's desk and historic equipment that had been used for important new discoveries. For a number of years our mutual visits were funded by the Volkswagen Foundation; there is a reason for this somewhat unusual source of finance. Nevill intended to build a new, modern Cavendish Laboratory and naturally a lot of money was needed to accomplish that. Therefore he asked the foundation whether it could contribute to the financing of this big project. The answer was negative, but instead a cooperation with a German university was suggested. Nevill asked me to contact the foundation and after my application it jointly gave us a grant which enabled our students and ourselves to make the necessary visits and it paid for a few technicians, too. In this way the stimulating and fruitful cooperation between the Cavendish Laboratory and the 'Marburg Group' began. In our many discussions I always admired Nevill's wide range of knowledge and his ability to pick out immediately the crucial point of a problem. Moreover, everybody who met him was impressed when at his age he didn't show any signs of tiredness, even after several hours of intense discussion.

There are some other events that I remember quite well. In 1968 Nevill attended a German conference in München, where I presented a review on

Celebrating news of the Nobel prize for physics with Professor Stuke.

amorphous semiconductors. Obviously he liked my talk because he invited me to present the initial paper at the Cambridge conference he organized in 1969. Due to this paper I became better known to the international community in our field and a number of new contacts arose. Therefore I feel that to a large part I owe it to him that at the Ann Arbor conference two years later I was asked to be chairman of the conference in 1973. Since Marburg didn't offer sufficient hotel capacity I decided to go to Garmisch-Partenkirchen. It was, of course, clear to me that Nevill ought to be represented in the programme by an important review paper, but unfortunately he couldn't attend the conference. He had a light pneumonia and his doctor advised him not to travel. I still remember the great disappointment of many participants and of myself when the central figure of our field was missing in Garmisch.

An event which I recall with somewhat mixed feelings is the celebration of Nevill's seventieth birthday. Abe Yoffe, who organized it, asked me to make a short speech at the banquet. My wife and I were overwhelmed by the historic splendour of the hall of Trinity College, where the banquet took place, and we greatly enjoyed the excellent meal; however, during the dinner I realized that I had prepared the wrong talk. Before me there spoke Heinz Henisch from Pensylvania State University and Brian Pippard, Nevill's successor as Cavendish Professor. Both told quite humorous stories of events happened during their long relations with Nevill, and the audience showed its amusement with smiles and laughs. In my talk, however, I only covered our scientific cooperation and I could easily see that the audience didn't like such a speech at this occasion and I became rather nervous. I ended my speech quoting an aphorism by G. C. Lichtenberg, the German physicist and philosopher living in Göttingen in the eighteenth century. Since Nevill spoke German well, I said it in German, but somebody from the audience asked me to translate it into English. I was unable to do that; I had forgotten to put the translation into my manuscript and I felt too nervous to translate the German version standing in front of the audience. Nevill must have been disappointed, but he didn't show it in any way. Since I feel that the aphorism describes his scientific personality quite well, I am glad to have an opportunity to repeat it here in my somewhat free English translation: 'It is peculiar that only extraordinary people make those discoveries which afterwards appear to be so easy and simple. It requires a deep knowledge to realize the simple but true relations between the things'.

For me, the most important of our meetings took place in early October 1977. Nevill represented the University of Cambridge at the five hundredth anniversary of the University of Tübingen, and after the celebration he and his wife came to Marburg and stayed at our home. Next morning Nevill and I started our discussions in my office; for lunch we walked, as we usually did, to the nearby restaurant Die Sonne in the centre of the old city not far from the town hall. We were joined by O. Madelung along with F. Hensel and W. Freyland from the Physical Chemistry Department. When we were just trying to find out what we should have as a dessert, the waiter told us that there was a phone call for 'a Professor Mott'. Nevill went to receive the call and we were wondering who might want to talk to him here, of all places, and why the call lasted so long. Since my wife knew that we would have lunch at this restaurant I thought that possibly something had happened to his wife, but after about ten minutes Nevill returned to the table smiling faintly and saying, 'I just learned that I got the Nobel prize.' He had been informed by an English journalist who had found out his presence at the restaurant via the secretary in Cambridge and my wife, then Nevill had given the journalist an interview. We were, of course, delighted to hear this marvellous news and toasted him with our wine. But soon Nevill wanted to be alone for a short while at this perhaps most important moment of his scientific life, so he took a walk. Madelung and I went back to my office, and after about half an hour Nevill returned. Then we started a celebration together with our students. We had champagne, flowers and some speeches, but we were quite often interrupted when Nevill had to answer phone calls, when journalists asked for an interview and when photographers wanted to take pictures. The photograph shows how happy Nevill was on this exciting and turbulent afternoon. Later on, some of my colleagues from the Physics Department and also the President of the University came to congratulate him, the day ended with a dinner party at our home in a more private atmosphere.

Next morning Nevill was interviewed by German television in our living-room; it was interesting to see how much time it took to record the interview compared with the five-minute broadcast in the evening. We had lunch with our wives, again in Die Sonne; during the meal surprisingly the Lord Mayor of Marburg come to our table, congratulated Nevill and handed over to him a medal of the city of Marburg. He had been informed by the keeper of the restaurant about our presence there and just walked over from the town hall. Nevill was quite delighted about this unexpected honour. In the afternoon he and his wife travelled back to Cambridge. In order to have more comfort on the long journey, their train and boat tickets were changed to first class, a benefit of the Nobel prize.

Besides this unique event there are two others I would like to mention, both of them in Marburg. In 1980 Nevill received an honorary degree from the Physics Department and at the celebration I gave the speech of honour; when I retired in October 1983 he made the main speech in my 'Festkolloqium'. The last time we came together was at the second Garmisch conference in 1991. Nevill arrived during the banquet and the applause at his appearance was enthusiastic. Since my wife and I were sitting close to him, we could see how, besides the Bavarian brass band and the folkloric presentations, he enjoyed the opportunity to meet friends from all over the world again. Next day he asked us to have dinner with him in his hotel and we spent a very pleasant evening together, not knowing that it would be the last time we met. Mainly due to a long serious illness of my wife and health problems of my own, our contact became restricted to the exchange of letters. The last one came shortly after his ninetieth birthday. I had sent him a congratulation letter from Baden-weiler, where I spent four weeks in a rehabilitation clinic. When I came home I found a long response from him; at the beginning he thanked me for my congratulations and mentioned our close relationship, but then he switched to a fairly extensive discussion of problems related to high-temperature supercon-ductors. This letter impressively shows how active he was, still working on science at his great age.

With the death of Nevill Mott there has departed one of this century's most eminent scientists; I consider it a great honour to have belonged among his friends.

JOSEF STUKE, physicist, was born in 1918 in Lastrup (Niedersachsen, Germany). He obtained his Doctorate (1947) at the University of Göttingen, and afterwards worked in industry on semiconductor research and development. In 1962 he returned to university at the Technische Hochschule Karlsruhe. From 1967 he was professor of experimental physics at the University of Marburg and he retired in 1983.

Swedish Involvements

Karl Berggren

The period I have in mind starts in 1970. By means of a scholarship from the Royal Society I then came to Cambridge as a postdoctoral fellow to work at the Cavendish Laboratory in Free School Lane. In my PhD dissertation, which I had completed about one year before coming to Cambridge, I had done work on the metal–insulator transition. From this point of view a stay at the Cavendish Laboratory thus appeared very attractive.

As I left Sweden I was told by my previous supervisor that I should, of course, not expect that I would ever get the opportunity of talking to Nevill Mott. He would be much too busy with more important things. I also had 'good advice' from other senior colleagues, comments like 'Why are you going to Cambridge? Mott is getting old and not doing anything these days.' In this way I got started on a long and rewarding interaction, scientifically and personally, with Nevill Mott.

As soon as I came to the Cavendish, I was given a desk in a room opposite Mott's and was invited into his office. We talked about general things, practical things about life in Cambridge, and also started on some physics. I still remember what an entirely new experience it was to talk to him. When I asked a question, he frequently remained silent for a long time, to the extent that I started to get nervous. Was my question that stupid, or did he not hear it? He had a way of disappearing into his own thoughts, and apparently forgetting about one's presence. Just as I had gained courage to repeat my question he returned his answer. A very precisely formulated and kind answer to exactly that question I had asked. I had an enormous feeling of being taken seriously. Mott had an informality and charm that made me forget differences in age (he was more than thirty years older than me) and status. Almost, anyway.

Informality and lack of prestige is an effective and truly intellectual way of communicating. If one is confident in oneself, like Nevill Mott, prestige and status become less important. Only essentials remain. This simplicity was also part of Nevill Mott's scientific style. He had a tremendous ability to look through a complex problem and structure it by making the right cuts. I came from a tradition in Sweden which was rather formal. One should be able to derive things or calculate things from first principles in a rigorous way. In Cambridge I was asked by Mott to estimate the Néel temperature associated with pairs of donors in silicon. The precise meaning of this concept is not important here, I just remember that I worked hard and did the best theoretical approach I could find. I got four degrees Kelvin, I think, and went to ask

31 SEDLEY TAYLOR ROAD
CAMBRIDGE CB2 2PN
Telephone 45380

There is a theoretical problem in the area of M-I branches that needs to be solved. Suppose one has a band-crossing Transition

and there is a discontinuous M-I transition when ΔE becomes less than the binding energy of the electron-hole condensed gas. Then at $T=0$ the free energy - c curve (c is any parameter, eg. composition

Letter from Nevill to Karl Berggren.

which varies ΔE, is thus

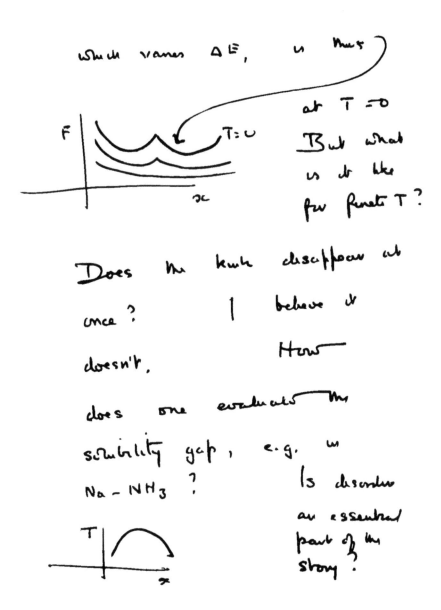

at $T = 0$

But what is it like for finite T?

Does the kink disappear at once? I believe it doesn't. How does one evaluate the solubility gap, e.g. in $Na - NH_3$? Is disorder an essential part of the story?

Letter—Continued

Mott what he thought. He had already done the estimate and believed it to be four degrees. Somewhat disappointed, I asked him how he could obtain it so quickly and accurately. Well, I felt it in my thumb, he answered. His type of intuitive physics was new to me, appealing in simplicity and elegance. The simplicity is deceptive, however; it is much too easy to end up with very weird answers. A colleague and friend of mine, Fumiko Yonezawa from Japan, later named it the Mott disease. To have the Mott disease is to pretend that one can dissect complex problems by making elegant and effective short cuts towards a conceptually simple model without having the intellectual insight and skill to do so, i.e., simply not being Mott.

After my postdoctoral stay in Cambridge I returned to Sweden in 1971, but not to my old institute in Stockholm. A new university was being set up in Linköping, a city some two hundred kilometres south of Stockholm. Thus I was invited to Linköping University as acting professor of theoretical physics while a number of applicants, including me, would fight for the chair on a more permanent basis. From a legal point of view the case was actually quite complicated. Indeed, there was fierce academic fighting, but this is not the place to discuss that. It meant, however, that I remained longer in Linköping than I ever anticipated. Starting out as the only theoretician in a situation like that, I ran the obvious risk of scientific isolation, too much teaching and administrative duties. Everything would have to start from scratch. Things were not that bad, however. In the first place, I could do whatever kind of theory I wanted, there was nobody around to push me. Secondly, there were letters. As many of us know, Nevill Mott had an enormous correspondence. His handwriting was distinctly personal and nice to look at, but was not always easy to decipher, as can be seen from the example shown here. In a number of letters he continued to discuss the nature of the metal–insulator transition and localisation and, as a consequence, the theory programme in Linköping was heavily influenced by his suggestions and enthusiasm. I was also very pleased that I could establish a programme in this way that was distinctly different from what one would find at other Swedish theory departments. Even more helpful were the visits that he made to our department and our emerging research group in condensed matter theory. I remember that I once lamented the lack of resources in Linköping and how I might envy him for his opportunities when he, for example, started his career in theoretical physics at Cambridge University in the early days of wave mechanics. 'Well, we were not that many either,' he replied. 'Among the theorists there were only me, Hartree and Dirac!' He did not say that to be funny or to ridicule. He was actually modest in his usual way and forgot that he and Dirac both became leading theorists of this century with a tremendous impact on modern physics and technology. At the same time, he was perhaps right: a big department is not necessarily a good department.

The visits to Linköping were not only about physics. Nevill Mott had a broad interest in education. He was curious about the Swedish comprehensive school. From working with various educational committees back in the UK, he claimed that the Swedish school system was regarded as very progressive among his colleagues. In a class every child would be able to study according to personal interests and talents. He therefore asked if we could visit a school, and so we did. At Österbergaskolan, a typical comprehensive school in Linköping, he attended lectures with 12-year-old kids, chatted with them and the teachers. As usual he made a quick contact especially with the children, who enjoyed him and his visit. Afterwards he joked that the visit to a Swedish school should afford him much prestige among his British colleagues in education. On the other hand, he was very questioning of what he had seen. To him, it all looked rather conventional. Why, for example, was there not a separate room in the back of the classroom to which pupils could withdraw for individual work. I tried to explain that perhaps not all teachers had been ready for or positive towards the reform to more progressive teaching that the state had just implemented in Sweden. 'How can one impose a reform without first making sure that all teachers are along and not ensure proper resources? It's only on paper that it looks fine!' he retorted. Thus he learnt about the Swedish comprehensive school, and so did I. I also had a lesson about sound pragmatism. I believe that I was a lot more ideological or dogmatic at that time.

Space is too brief to go into all discussions, interesting comments, etc. Here are some disconnected snapshots:

The Swedish Bofors anti-aircraft guns, I remember how we tested them for destroying barbed wire during the war.
Klaus Fuchs was the smartest theorist that worked with me in Bristol.
As soon as I learnt that electron distribution is the same as the wave function squared it was just to go ahead!
We are all very emotional about our theories.
[Talking about the Nobel prize.] I am beyond any thoughts of that, it does not really mean much to me now.

The problem with Nevill Mott, if it can be called a problem, is that he contributed so much, and did fundamental work in so many areas during his very long career. This did not make a selection more easy. Among the many topics, I believe he was nominated for work on the photographic process and dislocations. However, when the Nobel prize was later awarded to him, it was for the work related to disorder that he did during the years close to retirement. That too is exceptional. He came to Linköping after the ceremonies in Stockholm, talked to students and attended a party in the evening. He told about the television programme 'Geniuses speculate', a traditional TV interview with all Nobel laureates. Already in the studio, immediately after the interview, they could check how the programme had turned out. 'It was all very interesting,' he commented, 'but I cannot understand how they could develop the film so quickly!' It is a sweet comment from a genius who did not bother about the media world.

Enthusiastic letters from last years: 'I'm immersed in high temperature superconductivity.' 'The problem of the new superconductivity makes me feel rejuvenated. I feel as inspired as when I was a student at St John's with Dirac.'

KARL-FREDRIK BERGGREN is Professor of Theoretical Physics at Linköping University, Sweden. Since the early seventies he has maintained a close contact with research at the Cavendish Laboratory, especially with Nevill and Mike Pepper, and has spent several extended periods in Cambridge. His research interests are in metal–insulator transitions, semiconductor structures and devices, and electric transport in low-dimensional, ultra-small quantum systems.

My Life Touched by Nevill

Hellmut Fritzsche

It was extremely lucky for me that Nevill got interested in my work after I finished my dissertation in 1954 at Purdue University. Conduction phenomena in semiconductors at low temperatures and the metal–insulator transition were outside the field of vision for most solid-state physicists at that time. Nevill's interest was most encouraging when I joined the Institute for the Study of Metals of the University of Chicago in 1957 as it contrasted with that of my colleagues. Did they believe that I was working on metals? Perhaps they remembered B. Gudden's 1934 review article in which he had remarked, 'Metals such as silicon really should no longer be thrown in the same pot with semiconductors because the reason for a resemblance of their properties is entirely different' (my translation from the German).

Learning from Nevill the crucial role of the overlap of localized state wave functions, my students and I started to change their size and shape with uniaxial stress and improved the homogeneity and control over donor and acceptor distributions in germanium by nuclear transmutation doping. Nevill realized that exchanges of letters were not sufficient and sent W. D. Twose as a research associate to my laboratory for two years for a joint effort on the theory of impurity conduction.

Nevill was an active sponsor of the *Bulletin of the Atomic Scientist*, which was founded after the war at the University of Chicago to educate the public about the perils of the bomb and the nuclear age. I was more than delighted when Nevill visited Chicago and invited me to join his discussions with Cyril Smith, Harold Urey and Leo Szilard, who had just started his political action group, the Council for a Livable World.

Being in charge of our children when my wife, Sybille, took her bar examination in 1968 I took my 9-year-old son, Peter, to a Gordon Conference. He exclaimed, 'Professor Mott's lecture is the only one I understood,' which testifies to the power and clarity of Nevill's intuitive reasoning. On the other hand, I did not always find Nevill's letters to me easily comprehensible. Sometimes I had to read them several times, as Nevill left out several crucial steps of his argument which were obvious to him, not so to me. I should have consulted Peter but he had turned his interests to archaeology and history.

Gordon Conferences provide great opportunities for long walks and conversations. I was deeply touched when Nevill told me about his life as a youngster, which even then still gave him nightmares about dealing with the bullies in school, who had found many ways of making his life miserable.

Professor Fritzsche toasts Nevill on his eightieth birthday (1985).

Although he conceded that the strong scientific atmosphere in his parents' house had left him no choice but to study physics, we agreed that many scientists must have suffered in childhood, because they tended not to share the prevailing passions of their peers nor to conform to peer pressure.

On one of these walks, Nevill expressed his regret that solid-state physics had become the victim of computerized band structure calculations, pseudo-potentials and Green's functions and asked me to concur. I mentioned that I had met Stan Ovshinsky, who had opened for me a world of new problems in trying to make sense of noncrystalline disordered semiconductors. The downside, however, was that one got shunned by one's crystalline semiconductor colleagues. 'Then it must be important,' replied Nevill, 'I like to swim against the stream.' At the next opportunity, I introduced Stan Ovshinsky to Nevill, which was the beginning of a deep friendship, a long collaboration and a series of miniconferences with Nevill and a select group of disordered-state scientists at Energy Conversion Devices, Inc., in Troy, Michigan.

It was at times difficult to derail Nevill's attention from his preoccupation and focus him on new problems of one's own. I was deeply puzzled that my student S. Agarwal had found all chalcogenide glasses to be diamagnetic, even though the Fermi level appeared to be pinned by a rather large density of localized states. This seemed to be a fundamental problem and I explained my puzzlement to Nevill. After a long pause of deep thought, Nevill said, 'Do you think double injection might explain switching? What do we know about contacts?' Small consolation to know that I must have gotten through to him, because two years later he published the answer to my question.

Nevill loved to make fun of conventions and at the same time enjoyed the quaint traditions of Oxbridge colleges. Thus he put up a good show of absent-mindedness during one of our walks on the grounds of Caius College, when he edged further and further onto the immaculately groomed grass which he, as a Fellow, could step on but which I had to keep off. His dinner speeches were masterpieces of witty reflections, brilliant self-deprecations, so wonderfully British, and warm words of appreciation for the host and those present. Would the world never be without this, his, flair! Of course, one could detect a bit of British prejudice when the conversation touched on topics such as the Falkland Islands, the Royal family, class distinctions or English food, but Nevill was always willing to entertain a spirited counterargument.

When in the 1980s Nevill was again challenged by a big unknown, the puzzle of high-temperature superconductors, we would share some excitement and long discussions in his new office next to the Cavendish. My 1977 measurements with N. Sakai of pressure-induced superconductivity in an amorphous chalcogenide fitted well into his theory of superconductivity in lone-pair semiconductors. Moreover, Nevill was very intrigued by Ovshinsky's pairing mechanism in cuprates which was based on the local distortion accompanying the Cu^{II}–Cu^{III} valence change.

I will deeply miss a lifelong mentor and friend. Nevill has profoundly influenced my work, enriched my life and Sybille's too. He ennobled our field of science and taught us how much insight one can gain if one thinks – and listens to experimentalists.

Conquering a New Frontier

Sybille Fritzsche

London never seemed more cosmopolitan, more sophisticated, exciting and offering so many and varied cultural delights such as theatre, restaurants – and best of all a chance to escape our unruly brood of four children – than in August 1972. We had been travelling, first on the continent for six weeks, then for three weeks in Scotland with the four of them, ages 2 to 13, and beautiful and wondrous as the trip through the Western Highlands had been, it was also unrelieved 'togetherness' for a very long time in small confined spaces, i.e. a little car or an overcrowded bedroom. But here in London baby-sitters might be found, and better still, we discovered the Olympics were just taking place, broadcast twenty-four hours a day by the BBC. So we parked our children, negligent parents that we were, in front of the television while we dived into London's evening life, if not its night life. And the highlights of our adult exploits turned out to be a dinner with Nevill in the elegant but staid Athenaeum Club.

Remember, it was 1972 and, as they were called then, womens' libbers and their supporters among the menfolk had persuaded, by foul means or fair, the management of the club to integrate at least the dining-room. Few ladies had apparently availed themselves of this opportunity yet, and Nevill was eager to establish a precedent, taking the full measure of the as yet untried opportunity.

Hellmut and I met Nevill in the hall of the club after he had experienced the first of the pitfalls which such a bold course of action might present. I, obvious female that I was, could not enter the hall through the main door, oh no! There was a little, well-hidden side door through which the ladies had to proceed, and rather conveniently this helped me discover the whereabouts of the powder room. Well, that hurdle of obtaining entry overcome, we did meet and proceed into the dining-room. A great room it was, with wide open windows to which billowing curtains added a gracious note of summer charm; one looked into a lot of greenery, right there in the middle of London, rather surprising for those of us not well acquainted with the city's layout. And from the greenery there arose, just as surprisingly, the towers of Westminster Abbey and Big Ben. A splendid sight, a splendid table and a wonderful host, who proceeded to order good 'hock' as German white wine was referred to. We dined at leisure and well, interrupted by a clergyman, a bishop as Nevill informed us, who rose from a neighbouring table to ask me – very English-looking, as I shall describe shortly – for some advice on items on the menu. Hm, what was that about? He was obviously acquainted with Nevill and meant to show support by his bold action? Was he lending Nevill moral support as the waiters continued to serve us a bit nervously, and many glances, not all friendly, were cast in our direction? Or was it just to get a closer look at the *femme fatale*? Me, in my bright yellow Mexican evening dress, low cut, with bold green, orange and pink flowers on it? No discreet little black dress with the obligatory pearls for this 'coming out' of ladies in the sacred precincts of this distinguished men's club.

We had a great dinner, lots of good conversation which with more wine turned away from physics and physics gossip to topics of greater interest to me – and I do not mean law. We debated the merits of our favourite architecture: the Temple of Heaven versus the Pantheon, giving rise to rather lighthearted metaphysical speculations.

But all good things come to an end; the desert had been consumed with appreciation, now what? None of us thought of parting, but where to go from here? After all, Pall Mall is not exactly close to places catering to after-dinner delights.

Well, the success of the evening so far, and I venture to say the wine, had emboldened Nevill. We would push forward in our goal of fully integrating the club, so we marched behind him, with some apprehension, up and up the grand staircase to the second-floor library for coffee and brandy. And remember, dear reader, these were innocent times, cigars all round!

The library, as the name indicates, had quite a few imposing-looking leather-bound volumes, arrayed in dark, tall bookcases along the walls, interspersed with portraits of distinguished members of the club, lots of clergy, plenty of stern-faced and very properly dressed gentlemen in dark and somber suits, some heavily decorated uniforms, a rather imposing and rather stern array of portraits depicting superior males (Dead White Men would be the PC term of choice). And I swear that as we walked – slow pace – to the farther side of the library, there to sink into deep leather couches, the heads of the portraits turned around, staring at me! All of them! Well, finally we sank into those deep, deep couches, and Nevill, equally distinguished, both in achievements as in looks, glowered at the steward, who hesitantly approached the three of us to, I don't know what – to evict me, to lecture Nevill, to raise objections? The poor man had no chance to voice anything whatsoever. A little cough was firmly ignored and the order was given, 'Coffee, brandy and cigars all round.'

And as to the other, mostly ancient members of the club, sitting in the library. Well, one problem was the very comfort of those overstuffed and oversized couches and chairs: hard, with creaking bones and aching joints to rise from them, and with every minute where no objection was made, no firm steps taken to punish the transgressors, either much more in force, conviction and persuasion was needed, or the whole thing had to be forgotten. And knowing the British aversion to fuss and scandal in public or private places, there was a buzz, for sure, louder than normal in the library, I venture to say, but no action, no stern voices and no firm gestures evicting me from this sacred precinct.

Nevill had won – the last refuge of the English male was conquered!

PROFESSOR H. FRITZSCHE remained in contact with Sir Nevill since 1955, when he finished his dissertation on the metal–nonmental transition and impurity conduction in germanium at low temperatures. While he was professor at the University of Chicago his research associate W. D. Twose wrote a review on this topic with Sir Nevill, leading to a frequent exchange of ideas and spirited correspondence. In 1968 he introduced Sir Nevill to Stan Ovshinsky and the emerging field of amorphous semiconductors, which led to a close friendship and frequent visits of Sir Nevill to Stan's company, Energy Conversion Devices, Inc., in Troy, Michigan.

SYBILLE FRITZSCHE, Hellmut's wife, was staff counsel of the American Civil Liberties Union in the 1970s, director of the Chicago Lawyers' Committee for Equal Rights under Law in the 1980s, and became a Chinese historian in the 1990s. Nevill enjoyed lots of interesting discussions with her, and her contribution here records just one of their many unforgettable evenings.

Mott's Room

Stan Ovshinsky

Iris and I met and spent quite a bit of time with Nevill in San Francisco at the 1967 meeting of the American Physical Society. I recall an earlier brief introduction at a Gordon Conference I attended with Hellmut Fritzsche.

Several months earlier, in Leningrad, I had been invited to give a talk at a small gathering of primarily Russian and East European researchers, no more than thirty people. I was due to give one of the first scientific disclosures of my switching work. Kolomiets in his summary of the meeting drew an almost flat line on the board to show the progress in amorphous materials until this meeting and then said, 'With the work of Ovshinsky the field will now go like this,' and drew an almost vertical line.

That acceptance in Leningrad encouraged me to discuss our work with Nevill, Adler* and several others in San Francisco, and a group of us had a very nice dinner at a French restaurant where Hellmut and I carried on an intensive conversation with Nevill about the Ovonic memory, the Ovonic threshold switch and amorphous materials in general. He became very interested, albeit a bit skeptical that there could be such unusual effects. We therefore invited him to come and see for himself and arranged for him to visit our laboratory on his way back to England.

When he saw the devices actually working, he said that this work was very important, became quite enthusiastic and asked what he could do to help. This started a very warm and constructive working and personal relationship that continued until his death. He often said that our meeting led to his serious attention and contributions to the field of amorphous and disordered materials. He strongly supported my 1968 publication on switching in *Physical Review Letters* when it came under attack. He focused his powerful creative mind on the problems and possibilities of the amorphous and disordered field, made many important contributions, and working together with Ted Davis and others, he became its acknowledged leader.

We missed him at the international meeting on amorphous materials that fall in Bucharest; we were told he was unable to come as his father was very ill. He spoke about our work at various meetings and invited us to discuss it

---

* I had arranged to meet Dave Adler at the meeting because of my work in the 1950s with amorphous mixed-valence transition metal oxides which resulted in switching and memory devices. In 1960 I developed amorphous chalcogenides to demonstrate unique switching and memory devices, and they are the ones that we have commercialized.

and demonstrate our devices at the Cavendish Laboratory at the next international amorphous meeting that was held in Cambridge in 1969. This was a memorable occasion for two reasons. It was a wonderful experience to show our work at the Cavendish Laboratory. Our friendship became much deeper as he shared with me his personal feelings and the scientific and political frustrations that he had. His advice and encouragement were heartwarming. This closeness continued until the very end.

One of our great disappointments is that we were not able to set up a plant near the Cavendish, as Nevill had suggested, so that we could work more closely together. He introduced me to several university officials and we discussed plans of how an industrial park could be set up. One did come into being, but, unfortunately, we were not able to participate due our not having the financial wherewithall at the time. One of Mott's characteristics I admired was that he considered the technology that ensued from scientific discovery and activity should be put to everyday use to solve important problems in society. He wanted assurance that photovoltaics could be a realistic approach in Great Britain and was very interested in science policy.

The mechanism for the Ovonic threshold switch always intrigued him. Indeed, he was still discussing this with me in his last letter, shortly before he died. He had always believed it to be electronic in nature. He wrote several papers on it and was very pleased that experiments by various groups had proven this. The mechanism that I suggested was based on the fact that the chalcogenides, as pointed out by Kastner, were lone-pair materials. I suggested that the lone pairs were the basis for the unique electronic activity since they are a source of nonbonded electrons and, as such, they can respond reversibly to a suitable electric field without affecting the bonding electrons, which are much deeper in energy and responsible for the structural integrity of the material. The excitation process permitted injection from the electrodes and produced exceedingly fast switching. As long as the highly dense conducting plasma of carriers was sustained, the switch would remain in the On condition. Recombination of the carriers took place when the holding voltage was reduced. The Ovonic memory switches were designed through their bonding energies to permit a reversible phase change in response to the switching mechanism I have described. Nevill was delighted that the optical version of the phase-change switch (PD) had gained widespread commercial application.

We had ongoing scientific discussions about superconductivity. He had developed a polaronic model and I argued for a new class of bosons whose existence arose from changes of chemical structure in response to changes in local charge. It was very gratifying to see his great mind always at work with the deepest problems. He was particularly interested that we could apply the principles of disorder across a wide spectrum of usage: batteries for electric cars, photovoltaics, information systems, etc.

Whenever he came to the United States, which in the earlier days was once or twice a year, he would visit us and stay at our home in what we affectionately called Mott's room. When he stayed with us at the Ann Arbor meeting in 1971, he made breakfast for me, since I had never learned to cook and Iris suddenly had to go to the hospital.

When he visited our working group – including Hellmut Fritzsche, Dave Adler, Artie Bienenstock, Marc Kastner, Heinz Henisch and Morrell Cohen – all of us would meet and there would be much discussion and debate. On our many trips to Europe, we often visited him in Cambridge for a few days.

One of our most memorable scientific meetings was in 1976, where the Kastner–Adler–Fritzsche (KAF) model had its inception.

He came for the inauguration of our Institute for Amorphous Studies, as a member of the international advisory committee, and he gave one of the early lectures there. That year was a good one for octogenarians at our Institute, three lecturers in a row: Mott, Linus Pauling and Doc Edgerton of MIT! On behalf of the Institute, Hellmut Fritzsche, David Adler and I presented him in Cambridge with a three-volume festschrift for his eightieth birthday.

He was greatly interested in our work in photovoltaics and urged me to submit a paper to *Nature*. Published in 1978, it showed how the physical properties and the density of states in amorphous silicon alloys could be minimized and controlled.

He kept up a continuous correspondence with me to the end. His handwritten letters came in envelopes similar to wartime V-mail and besides his personal comments about what was going on in the various fields of science, they always concentrated on the scientific problems at hand, which later

A meeting held at Energy Conversion Devices: Stan Ovshinsky is seated on Nevill's right.

David Adler, Hellmut Fritzsche, Stan Ovshinsky, Nevill Mott and Iris Ovshinsky.

included high temperature superconductivity. In his letter writing, he was of the old English school, always informative and stimulating, and his mind never flagged at attacking the most difficult problems. The richness and depth of his scientific thinking was always apparent.

We celebrated many of his birthdays with him and Lady Mott. Iris and I were deeply appreciative of what he said at our last meeting, his ninetieth birthday celebration, where he acknowledged with great generosity our influence on him in the scientific sphere. We have lost a wonderful friend who was a model of how one could age and yet be ageless in one's interests and important contributions. He was truly a great man.

MR S. R. OVSHINSKY is President of Energy Conversion Devices (ECD), a company he founded to develop and exploit the unique properties of amorphous and disordered materials. His wife, Iris, works with him and shares his scientific interests. Together they met Nevill on numerous occasions in Cambridge, at ECD and at many locations (often conferences venues) around the world, maintaining a close friendship for thirty years.

My Research with Sir Nevill

Moshe Kaveh

My First Acquaintance with Mott

I first came to Cambridge in 1979 to spend a sabbatical year at the Cavendish Laboratory, and I naturally sought out a research collaborator. Even in my very first conversations with Mott, I felt a strong 'chemical bond' quickly forming between us.

Regarding which lines of research to pursue, Mott told me, 'Moshe, I recommend that you *not* work on the metal–insulator transition because that area of physics is all played out.' A scant few weeks later, the Gang of Four published their famous article, showing that the Anderson transition was continuous and not of first order. Mott called me over to say, 'You see, Moshe, this proves that not everything I say is correct.'

At almost the same time, Altshuler and Aronov published an article on the influence of interactions on metallic conductivity. Mott asked me to explain their article to him without using any diagrams. This indicated to me how he thought about physical problems. Our initial work together, ultimately resulting in two joint papers in 1981 on the logarithmic dependence of the metallic conductivity in a two-dimensional system, forged the strong bond between us that lasted until his death.

Over the years, I published dozens of articles with Mott, with each joint paper serving to strengthen and deepen our relationship. I came to see him as my spiritual and moral mentor. Mott particularly impressed me with his completely objective approach to science. He never cared who was the originator of an idea, whether student or Nobel laureate! It was always the *idea* that counted with him. Mott had a cheerfulness and simplicity about him – and a passion for science. He also had clear and definite views on most topics of interest, which he often expressed in writing, courageously defending his views even when they were none too popular.

Mott, the Scientist

Mott had a unique style of thought for a theoretician, always seeking a simple physical argument that could provide a short cut through complex mathematics. In our many joint projects, he often took home with him over the weekend the detailed mathematical analysis that I had worked out and returned with physical arguments that yielded almost the same result. He would appear at the lab on Monday morning with a shy smile on his face, and say that he thought my detailed analysis was correct because he had succeeded in obtaining the same answer, up to a factor of π, using only very simple ideas! Mott had an enormous respect for experimental results, in complete contrast to his initially sceptical attitude to any new theory.

Although he made major contributions in so many diverse fields, Mott once told me that he received his greatest pleasure from his work on disordered systems and the metal–insulator transition.

Mott unrelentingly sought the truth, and would often critically examine articles that he himself had published many years earlier, seeking alternative approaches to help him explain the problem that was troubling him at the moment. He never feared making a mistake, always emphasising that one can learn as much from an error as from a new idea. When he did err, he admitted it openly and in writing, with a candour that can only be admired.

I was particularly impressed by his ability to approach a topic and, within a few hours, to summarise and explain the salient features in an original way in the form of an article which, after a week or two of polishing and minor corrections, was ready for publication. The pace of his writing was astonishing! When he decided to write a book, he would work with enormous energy and speed, all *without* using a computer.

His tendency toward brevity in writing enabled Mott to reply *personally* with a handwritten answer to every letter he received, be it from a Nobel prizewinner or from a beginning student in physics. I have a collection of literally thousands of his letters to me, filled with graphs and equations, but he always ended each letter with some warm personal remarks. Because he answered every letter that he received, he was sent an enormous number of preprints. His office was a sort of library of articles from all over the world, about which he had expressed an opinion. He kept the articles he read in a file, that could be called Mott's personal library. I would mentally separate all the articles he received into two groups: those that Mott considered worthy of being placed into his file for weekend reading, and those that were not deemed so worthy.

Mott, the Person

I spent every summer in Cambridge for seventeen years, working with Mott. In addition to the thrilling scientific adventure of carrying out research with such an intellectual giant, I was profoundly impressed with Mott as a man with a multifaceted personality. Let me relate a few anecdotes.

Mott had a fine sense of humour and loved a good joke. One Sunday, he invited me and my wife to his home for tea. When I asked him what time to come, he emphasised that he was not one of those typical Englishmen who insisted that Sunday tea must be taken at exactly 4:00 P.M. But when I indicated that I did need to know approximately when to arrive, he replied, 'Since I'm not the typical Englishman, come at 4:15 P.M.'

In another incident, he apologised to me and my wife that his garden was not properly cared for, explaining how he felt uncomfortable in requiring his elderly gardener to tend it more diligently. When I asked just how old the gardener was, Mott replied that he was already 77. Mott himself was then 75!

Mott would arrive at the Cavendish Laboratory early in the morning, and leave at 5:30 P.M. One day, as he was leaving, he passed by my door and asked if I wanted a lift home. I replied that I planned to stay a while longer in order to finish some mathematical analysis. He then told me the following story about Lord Rutherford.

Rutherford would always leave the Cavendish Laboratory promptly at 5:30 P.M. One day, before leaving for home, Rutherford passed by the laboratory of James Chadwick (later Nobel laureate, but then his research student) and asked if he could get a ride home with him. Chadwick replied that he planned to stay longer in the laboratory because he was very busy trying to finish his experiment. Rutherford therefore went home alone and, after supper, as was his habit, he went for a stroll in what is now Rutherford Lane. As Rutherford passed the Cavendish Laboratory, he saw his student still engrossed in his work. Rutherford pointed out to Chadwick that it was already 7:00 P.M. and he had not yet gone home. Chadwick answered, 'Professor Rutherford, so much work has piled up that I must stay even longer to

Nevill in 1983 with Professor Moshe Kaveh at Bar-Ilan University, Israel; during his visit Nevill received an honorary degree.

finish.' Rutherford then replied with the saying that Mott loved so much, 'Chadwick, if you spend so much time working so hard, when will you have time to think?' Mott added that Chadwick took this advice to heart, and eventually won the Nobel prize!

As is well known, many scientists would visit Mott. Among his numerous visitors, Mott had little interest in the abstract mathematicians, whom Mott described as people with answers who were looking for a suitable question!

Once Mott asked me to join a meeting with four eminent mathematical physicists who came to see him. After an hour-long presentation of their work, Mott told me he did not understand a single word they had said. When I asked Mott why he had not commented during their presentation that he wasn't understanding them, he replied, 'I didn't want to embarrass them by making them feel that there was no value to what they were doing.'

Mott had an unusual approach to contending with unusual situations. About twenty years ago, he started to take an interest in religion, and every summer he would question me at length about Judaism, its laws and its rituals. He asked me for literature about Judaism, and after reading these items, he returned to me with the following statement: 'You Jews must be very smart to understand all these laws and rituals. Your religion seems even more complicated than disordered systems. Therefore, the information transfer from Judaism to the Jew must be a first-order phase transformation!'

Because of Mott's great interest in Judaism, over the years, we had many theological arguments and discussions about comparisons between Judaism and Christianity. On one occasion, I asked him to speak to the London chapter of Friends of Bar-Ilan University, a fundraising organisation for the university. Mott chose as his topic 'The contribution of Jews to science'. After his lecture, a well-known London Jew told me that he had not realised Mott was Jewish!

Mott was once invited by the Church to deliver a lecture on religion and miracles. Mott's main thesis was that miracles, supernatural events, simply *do not exist*. As one can imagine, his lecture generated considerable antagonism in the Church. When he returned to the Cavendish on Monday, I asked him how his lecture had been received. Mott replied in his usual matter-of-fact tone, 'I don't think that I convinced them.' When I asked him, 'And what will you do now?' Mott responded that he would write a book on the subject! Thus, came to be born his book *Can Scientists Believe?*

When I first arrived at the Cavendish Laboratory, I asked the where-abouts of Mott's office. I was told, 'It's in the Mott Building!' I once asked Mott if it didn't embarrass him to work in a building that bears his own name. He replied with his own special brand of humour, 'When I was younger, Moshe, it did in fact embarrass me, but now I find that it helps me to locate my office!'

Mott had an almost mystical belief in the principle of free will, and he wrote several articles on the subject. His most famous saying on this was, 'In spite of my unshakable belief in quantum mechanics, it is difficult for me to accept the fact that the human brain is nothing but a wave function.' This most certainly applied to Mott himself! Science has lost a towering figure and I have lost a dear friend. His personality will forever remain engraved in my memory.

PROFESSOR M. KAVEH is currently President of Bar-Ilan University where he has been Professor of Theoretical Physics for over twenty years. His research interests in condensed matter physics extend over the fields of disordered systems, localization, metal-insulator transitions, quasi-one-dimensional systems, mesoscopic and statistical physics. He collaborated extensively with Sir Nevill while holding a Royal Society Visiting Professorship at the Cavendish and subsequently during many visits.

Personal Memoires

Hiroshi Kamimura

During the forty-two years of my academic career, I have on innumerable occasions been indebted to senior people, friends and colleagues. Above all, it is the long acquaintanceship with Sir Nevill Mott to which I owe most, and which I cannot fail to acknowledge.

In 1973, following a letter I wrote to Professor Mott (whom I did not know at that time) asking him for opinions on a problem concerning Anderson localization, he replied with such friendliness that it seemed as if we had known each other for a long time. This communication triggered his eventual invitation in the following year for me to visit the Cavendish Laboratory as a guest researcher for nearly one year, starting in October 1974. With this visit came the honour of working in collaboration with Professor Mott, and of being the first Japanese scientist to do so.

When I was an undergraduate student at the Physics Department of the University of Tokyo in 1953–54, the textbooks written by Professor Mott such as *The Theory of Atomic Collision* (Mott and Massey), *Electronic Processes in Ionic Crystals* (Mott and Gurney), *Theory of Metals and Alloys* (Mott and Jones), etc., were very popular among senior undergraduate and graduate students, postdocs and researchers. I myself learnt the great pleasure of studying solid-state physics by reading the Mott and Jones textbook when I was a third-year undergraduate student. At that time, we physics students in Japan knew these textbooks were written when Professor Mott was in his thirties. It thus became natural for Japanese physics students in the 1940s and the 1950s to look upon him as a god in physics.

Yet eminent physicist of the highest calibre that he was, he did not in the least stand on ceremony. He was very kind and generous to everyone. During my stay in the Cavendish Laboratory from October 1974 to August 1975, he was very thoughtful to adjust his schedule so that we could meet almost daily in his office in the Mott Building or in my office in the Bragg Building, or at tea time or lunch to discuss common problems in physics. The first subject we concentrated on was the mechanism of the ESR-induced decrease in the resistance of doped silicon and germanium in the Anderson-localized concentration region, as observed by Professor Kazuo Morigaki. After we wrote a paper, Professor Mott wanted to publish it in *J. Phys. Soc. Jpn* (*JPSJ*). This, he said, was because it was his first paper with a Japanese coauthor and many interesting experimental results on Anderson-localization had been published in that journal. Our paper 'The variable-range hopping induced by electron spin resonance in n-type silicon and germanium' appeared in JPSJ, Vol. 40, No. 5, pp. 1351–58 in 1976.

During my 1974–75 stay in Cambridge, along with my family, the constant care offered us by Professor and Lady Mott ensured it was the most enjoyable of visits. One day they invited my daughter and son to their house. There our children were pleased to play with Nevill's grandchildren. And my son, who was 9 years old at that time, had fun with the building blocks Professor Mott had enjoyed during his own childhood around 1910. Professor and Lady Mott kindly offered to donate a portion of the building blocks to my son. These building blocks, which retain a flavour of British culture at the beginning of the twentieth century, are now a treasure of my home. Nowadays my grandson enjoys playing with them.

After I worked with Professor Mott in 1974–75, whenever the opportunity arose, I paid him a visit. As a result, I met him almost every year until 1995. It was always a great pleasure for me to see and talk with him. I would like to tell some of my pleasant experiences.

In 1976 Professor Mott and I were chosen as plenary speakers at the Thirteenth International Conference on the Physics of Semiconductors, held in Rome during August 1976. I still remember that we stayed in the same hotel, Hotel Nazionale, and that Professor Mott kindly took me to several historical places around the hotel, such as the church Santa Maria Maggiore, acting as a guide by translating Italian explanations into English.

In 1978 Professor Mott invited me as one of the lecturers at the Nineteenth Scottish Universities Summer School on Metal–Non-metal Transitions, held in St Andrews. In my heart I still cherish the memory of three wonderful weeks I spent with Professor Mott and other participants at St Andrews and his generous hospitality.

In 1989 Oxford University Press published my book *The Physics of Interacting Electrons in Disordered Systems*, written with Professor Hideo Aoki of the University of Tokyo, my former graduate student. We were not only privileged to have the preface kindly penned by Professor Mott, but were also honoured with various kinds of advice he offered, our gratitude for which we cannot fully express.

Through these and similar occasions my awe and respect for a great man have increased every year. But these feelings are by no means unique; they seem common to all who met him. A number of Japanese researchers had received good correspondence from him. As a result, Professor Mott was very popular in the physics community of Japan and many Japanese researchers had visited the Cavendish Laboratory to see him. We call this phenomenon the Mott syndrome. During my stay in the Cavendish Laboratory I saw that Mott syndrome affects many people throughout the world.

In 1990 I was asked to write about the Mott syndrome by the *Nihon Keizai Shinbun* (the *Japan Economic Newspaper*, one of the country's four major titles). Written in Japanese, my article appeared in the 18 October edition for 1990, and this is how the name of Professor Mott became known

Hiroshi Kamimura (right) with Nevill outside the IRC in Superconductivity.

to business and industrial people in Japan. My article was kindly translated for Professor Mott by Dr David Ko, then a postdoctoral fellow in my group at the University of Tokyo, just before I retired. Professor Mott was very pleased to read the English translation.

Since 1990 I have organized the UK–Japan government-based collaboration programme on the subject of low-dimensional structure and devices (LDSD), whose projects successfully finished on 31 March 1995. During this programme I could visit Cambridge every year and stayed in the Interdisciplinary Research Centre in Superconductivity to discuss the mechanism of high-temperature superconductivity with Professor Mott, Sasha Alexandrov and Yao Liang, Drs John Loram and Joe Wheatley and other researchers. In particular, I enjoyed enlightening discussions with Professor Mott. When I gave a seminar on 3 March 1995, Professor Mott introduced me by surveying our long friendship since 1973. This was the last time I met him. His kind words at the introduction of my talk are still fresh in my memory.

Professor Mott was my great leader, not only in physics but also in my life. He was a man of the greatest personal kindness and generosity. It is a wonderful thing that I have been acquainted with him and through him have learnt an uncountable number of things.

PROFESSOR H. KAMIMURA was born in 1930 and obtained a doctorate of science in 1959 at the University of Tokyo. From 1961 to 1964 he worked at Bell Telephone Laboratories, Murray Hill, USA. Subsequently he was appointed lecturer and, in 1978, professor of physics at the University of Tokyo. In the 1980s he served as President of the Physical Society of Japan and Chairman of the IUPAP Semiconductor Commission. In 1991 he moved to the Science University of Tokyo, where he was Dean from 1993 to 1996. His research interests are in theoretical condensed matter physics and currently include high-temperature superconductivity, low-dimensional semiconductors and clusters.

Reminiscences

Kazuo Morigaki

The last time I saw Professor Sir Nevill Mott was in Cambridge in 1993 on the occasion of the Fifteenth International Conference on Amorphous Semiconductors (ICAS). I was really shocked and very sad when I heard about his death from a friend one day after it was reported in the Japanese newspaper *Asahi Shinbun* on 11 August. Until this moment, I had thought of him as being eternal, because when I saw him at Cambridge, he was well and still active for research.

I was a member of the International Advisory Committee of ICAS (previously called ICALS) which Mott chaired for a long time until 1981. I remember very well the committee meeting held at the ninth ICALS in Grenoble in 1981, where the site of the next ICALS in 1983 was discussed. Tokyo was nominated as a candidate and Mott supported our proposal. I was then the conference secretary, so I prepared the discussion documents. Without other proposals, Tokyo was chosen as the venue for the tenth ICALS.

I met Mott for the first time in September 1972, more than twenty-five years ago. At that time, I visited the Cavendish Laboratory to see Professor Ted Davis. As I did not expect to meet Mott, I was very excited when it happened. He kindly showed me the High Magnetic Field Laboratory (Mond Laboratory) where Kapitza had worked, and also the small museum, where he pointed out his father and mother in an early picture of graduate physics students in Cambridge. When I was an undergraduate student in the Department of Physics, Osaka University, I had a chance to listen to him lecture at the Osaka Gas Building Hall in 1953. He visited Japan to attend an international conference on theoretical physics held in Kyoto. At that time I knew his name from his book *Wave Mechanics and Its Applications*, coauthored with Dr I. N. Sneddon, with which I learned quantum mechanics on my undergraduate course. When I was invited to his home in Cambridge in 1974, I told him this story, which pleased him. In his home he said that his visit to Japan was very impressive. His daughter Alice told me that, after coming back from Japan, he talked daily about his journey. When he visited Sapporo, Hokkaido, many journalists and camera crew came to the platform of the railway station to meet him, which had surprised him. In Cambridge I told him about a future project to construct a tunnel under the Tsugaru Straits, to connect Hokkaido and Honsyu (mainland). He said that, although this would undoubtedly bring more convenient transport, it might cause destruction of the natural environment. He cited a similar problem on a proposed tunnel under the English Channel. I think he was always concerned with the environment.

Kazuo Morigaki and Nevill at the banquet of ICAS 14 in Garmisch-Partenkirchen (1991).

I visited Mott at Cambridge during my stay at the University of Hull in 1977–78. I was invited by him to stay in Caius College, and also to participate in a dinner held there in honour of Professor Hans Bethe on the occasion of his visit to Cambridge. I spent a very exciting time with Professor Bethe and his wife and other distinguished guests in the Mott party. After the dinner, I visited the Cavendish Laboratory together with Bethe. I was very interested to hear Mott's answer when Bethe asked him which research led to his Nobel prize. Mott said that it was for his work on amorphous semiconductors.

Mott was open-minded when he discussed physics with young researchers, so that he was respected and loved by many experimentalists and theoreticians. People having an exciting discussion with him for the first time were fascinated by his generous nature. This was named the Mott disease by Professor Fumiko Yonezawa at the banquet of the seventh ICALS, held in Edinburgh during 1977 and attended by Mott.

Discussions with Mott were vivid and suggestive on every occasion. I remember well discussions with him in 1974 when I visited him in Cambridge. At that time, I was engaged in spin-dependent conductivity experiments on n-type crystalline silicon, concerning the electronic states around the metal–insulator transition concentration of donors. I had looked at the resistivity change associated with electron spin resonance (ESR) in phosphorus-doped silicon. ESR led to a decrease in the resistivity for donor concentrations on the insulating side. This was accounted for in terms of spin-energy transfer from the donors to the electrons responsible for electrical conduction. Mott suggested to me that conduction might be due to variable-range hopping, so that hopping electrons should be responsible for the ESR. His idea was published in a paper coauthored with Professor Hiroshi Kamimura. Our discussion was expected to continue in the afternoon, but Mott left a memorandum for me, because he had to leave on urgent business. The memorandum expressed his idea from another viewpoint and gave an equation concerning the temperature dependence of the mobility. But I could not understand the meaning of this equation, so I asked Ted Davis when I met him at the International Conference on the Physics of Semiconductors held in Stuttgart just after my visit to Cambridge. Ted said that nobody understood it; it was just an inspiration of Mott. In his papers one often meets equations derived by physical insight. His greatness is that the results he derived intuitively expressed the essence of physical phenomena. An equation expressing the temperature dependence of hopping conduction, i.e. Mott's $T^{-1/4}$ law, is one example of this. He attempted to understand phenomena simply and physically.

He had recently devoted himself to high-temperature superconductivity, but I think he always thought about amorphous semiconductors. We were very pleased to have his message at our amorphous semiconductor conference, the sixteenth ICAS, held at Kobe in 1995. His message was sent through Professor Ted Davis; here is a part of it:

I have been engaged in the last few years with the study of high-T_c superconductivity, which is really an extension of what we studied a decade ago. Conduction in impurity bands in Si:P was an example of movement in a non-crystalline environment. We find that if our semiconductor is a Mott insulator or two-dimensional, the carriers must form polarons. These form charged bipolarons, and hence are superconductors. May I recommend the subject for people who understand impurity conduction.

This was an important message for me in relation to my work on impurity conduction.

Finally, I wish to mourn Professor Sir Nevill Mott's departure and to express my heartfelt gratitude for his warm and generous support of my research on the metal–insulator transition and amorphous semiconductors over a quarter of a century.

PROFESSOR K. MORIGAKI was born in 1932 and obtained his doctorate of science from Osaka University in 1959. He is currently Emeritus Professor of the University of Tokyo and Professor in the Department of Electrical Engineering at the Hiroshima Institute of Technology.

Growing up with Sir Nevill

Tiruppatur Ramakrishnan

I first came across Sir Nevill Mott in 1960, through his little book on wave mechanics, in a dusty bookshelf of the Central Library of Banaras Hindu University, Varanasi. At that time, quantum mechanics was mainly taught in the Mathematics Department of the university, and the refreshingly physical approach of the book was an eye-opener. His more difficult book, written with Massey, revealed him to be an atomic physicist. I was at that time interested in nuclear physics, and we peered through microscopes for hours at tracks of high-energy particles in photographic films. How these blobs of silver form and stay was a question which puzzled me; one of my teachers told me that the best place to start was yet another book, Mott and Gurney! With this early background, I was prepared to see Mott everywhere, and as it happened, I went into the field where his many ideas are central and critically influential.

Along with many others, I continue to think about some of the questions Mott raised, the answers he gave, and the supporting evidence he put together, all in his uniquely direct and inclusive way. It is humbling to realize that more than fifty years after its initial articulation, the interaction-driven metal–insulator transition is not fully understood. Is it a discontinuous transition connected with the unscreening of Coulomb interactions? Or is it localization due to short-range correlation in a commensurately occupied lattice? Or does inevitable static disorder pre-empt both these limiting possibilities? All these ideas have been put forward by Mott, who developed many celebrated and unreasonably effective criteria describing their occurrence. None of these phenomena is fully understood, either singly or in combination (in real systems they often occur in combination).

Like a great and thoughtful explorer, Mott has left behind first-hand accounts of the new territories he charted. The guidebooks for which I am particularly grateful describe metal insulator transitions and electronic processes in non-crystalline materials. Their sweep and deceptive simplicity amaze me still.

I did not know Sir Nevill very well, having met him and corresponded with him only occasionally. I cherish his handwritten letters. One in particular came at a time when I had moved to Varanasi, nearly twelve years ago, and was feeling somewhat isolated. He made generally encouraging remarks, and had thought-provoking questions on how magnetic fields or strong inelastic processes could cut off localization, and how this was manifested experimentally in several systems he mentioned. It was a timely tonic, from a great doctor.

PROFESSOR T. V. RAMAKRISHNAN took his first degree at Banaras Hindu University, Varanasi, India, and did his PhD at Columbia University, New York, USA. He has worked in several universities and institutes in India and the United States. He is a Professor of Physics at the Indian Institute of Science, Bangalore, India. His interests are in the physics of disordered and strongly interactive electronic systems, high-temperature superconductivity, as well as in the liquid–solid transition and related phenomena.

Fond Remembrances

Ram Rao

I first got to know Professor Mott personally in 1974. He had come to Oxford to give a seminar at the Inorganic Chemistry Laboratory (ICL). The topic of his talk was the effect of disorder on the transport properties of solids. I was then a Commonwealth Visiting Professor at ICL. I met him soon afterwards and later wrote to him at Cambridge, describing results on some of the interesting metal–insulator transitions in perovskite oxides that I had obtained. The transport properties of these materials seemed to conform to those expected in disordered systems involving variable-range hopping at low temperatures. He got very much interested and asked me to visit him at Cambridge. I went by bus from Oxford to Cambridge. To my surprise, Professor Mott was at the bus station in Cambridge to receive me and we spent an entire day talking about the results. A couple of weeks later, he visited Oxford and spent considerable time with me. I remember, during one such visit, when he was in my room at the ICL, there was a frantic call from the Vice-Chancellor's Office looking for Professor Mott. Apparently, he was to receive, within the next few minutes, an *honoris causa* DSc from Oxford, but he was busy talking with me about oxides and transport properties. When I visited him later in Cambridge, I found the diploma of the DSc degree from Oxford on a noticeboard for others to see.

I wrote a paper with Professor Mott in early 1975 and returned to India soon after, but I maintained correspondence with him. A few years later, I had completed some work on certain insulator–metal transitions in perovskite oxides, showing insulating behaviour at low temperatures, with a curious temperature dependence. Professor Mott was interested in this behaviour and included it in an article in *Advances in Physics*.

My association with Professor Mott became much closer when I came to Cambridge as a Nehru Professor in 1983–84. I used to meet him at the Cavendish. He enjoyed social evenings. Once or twice, he suggested that my friend Peter Edwards should arrange a nice party in one of the colleges. He would come quite early, have one or two aperitifs, enjoy the meal with good wine and have a cognac or brandy later. Around 9:30 or 10:00 P.M., he would say, 'Well, I think I should start driving home.' The drive was to Milton Keynes. I always wondered how he could do that at the age of 80. I cherish those evenings and some of the hours I spent with him talking about his younger days in Cambridge and also about his views on various aspects of condensed matter physics and science as a whole.

I found Professor Mott to be a shy, fine person. To start with, I was a bit worried as to how he would react to a chemist like me being conscious of the eminent position he held in the field of physics. He was, however, most charming and enjoyed learning about complex oxides and other systems. He did not hesitate to ask an elementary question on some chemical aspect. In 1985, when I was in Cambridge, Peter Edwards and I edited *The Metallic and Non-Metallic States of Matter* (published by Taylor & Francis) to celebrate Professor Mott's eightieth birth anniversary.

After 1985 I used to visit Cambridge occasionally, and whenever possible I would go to see Professor Mott. He seemed excited about high-temperature superconductivity and the cuprate materials. It always touches me to remember how he would drive all the way from Milton Keynes to see me at Cambridge. I used to worry about him driving such a long distance, but he did not seem to mind. It is difficult to find a person like him who had so much interest in his science and in his collaborators.

In 1994 there was a special issue of the *Journal of Solid-State Chemistry* for my sixtieth birthday and, lo and behold, the first article was by Professor Mott, where he even referred to some of my work on superconducting cuprates. I was flattered by this gesture.

It was indeed a matter of great pleasure that I could edit, along with Peter Edwards, a volume entitled *Metal Insulator Transitions Revisited* for the ninetieth birthday of Professor Mott. This volume was published by Taylor & Francis in 1995.

In early March 1997 I helped to organize a Royal Society discussion meeting on metal–insulator transitions. Professor Mott had wanted to attend the meeting and make some observations, but it was not to be.

Professor Mott was a statesman of modern condensed matter science. His simplicity, his childlike curiosity and extraordinary commitment to science were exemplary. There is much more one can say about him, but most important of all, I feel that most of us who have known him personally feel that we have lost a very dear friend and a well-wisher.

PROFESSOR C. N. R. RAO is Albert Einstein Research Professor and President of the Jawaharlal Nehru Centre for Advanced Scientific Research in Bangalore, India. He is a Fellow of the Royal Society and Foreign Associate of the US National Academy of Sciences. His research interests are in solid-state chemistry.

Science and Social Life with Nevill in Grenoble

Claire Schlenker

I met Nevill Mott for the first time when I was a young postdoctoral scientist working for a few years only on the metal–insulator transition in transition metal oxides. This was during the bicentennial French–British Conference jointly organized by the Institute of Physics and the Société Française de Physique in 1974 in Jersey. My colleagues and myself were naturally very impressed by his stature, both scientific and physical. We greatly enjoyed discussing with him, not only because of his great knowledge of the subject, but also for his considerable interest in modest but new experimental results.

Somewhat later, in 1976, he was one of the most prestigious participants in the CNRS Meeting on the Metal–Nonmetal Transition organized under the leadership of B. K. Chakraverty in Autrans, near Grenoble. Close contacts were then established between our laboratory and Sir Nevill Mott. He was several times our guest when B. K. Chakraverty was directing the lab and after 1984 when I was the director. Nevill came either as an official member of our external scientific committee or to take part in informal discussions as an efficient adviser. He also was made doctor *honoris causa* of the University of Grenoble during that period. The most striking memory I keep from his visits is related to his ability to devote the same interest to the data obtained by the youngest and most obscure physicist as by the most senior one. He could thus give strong encouragement to everybody, especially young people. Of all the high-level scientists I have met, I believe he was the most stimulating towards new developments. After listening carefully, he used to go to the blackboard and give his comments and suggestions with great brio.

Science with Nevill Mott was conducted in English. But when we entertained him at home or in a restaurant during the evening, he would start to speak French. After a long day of scientific discussions, he was able not only to take part in various conversations in French, but also to tell us fascinating stories about his professional travels all over the world. He was still able to do that during what I believe was his last visit to Grenoble in 1987, when he was 82. My colleagues and myself were amazed by his talents as well as by his energy. I should add that Nevill was a delightful guest also for his ability to enjoy good food and drink!

I met Nevill Mott a number of times during my scientific life. I happened to be in Washington at the 1978 March Meeting of the American Physical Society, where he gave a special talk for his Nobel prize. I enjoyed the small meeting on magnetite and related compounds he held in Cambridge in 1979. I

Dining *en famille* with Professor Claire Schlenker in Grenoble (1987).

took part in the 1983 NATO Advanced Study Institute on Physics and Chemistry of Electrons and Ions in Solids, coorganized by Nevill in the beautiful college of Gonville and Caius, Cambridge. This was a memorable meeting with a very friendly atmosphere, due to his exceptional personality. I always enjoyed talking with Nevill and believe I took advantage of his advice. I am sure he had a great influence on my way of doing physics, as he probably influenced many colleagues.

As a conclusion, I would like to say that Nevill Mott was a Grand Monsieur not only in science, but also in other aspects of life!

CLAIRE SCHLENKER is Professor of Physics at Institut National Polytechnique de Grenoble (INPG) and belongs to Laboratoire d'Etudes des Propriétés Electroniques du Solide, CNRS, Grenoble. She was a former director of this laboratory and is now Director of Ecole Nationale Supérieure de Physique de Grenoble at INPG. Her research fields include electronic instabilities in transition metal oxides and bronzes, metal–insulator transitions and charge-density waves.

Beyond the Mott Transition

Yao Liang

Nevill was a giant in physics. In 1963 I joined the Cavendish Laboratory where Nevill was Head of Department to study for a PhD under Abe Yoffe. It appeared to me that I had come to a place so full of great physicists that it was almost unreal, but Nevill seemed to move on a higher plane – above all the others. It is often said that he was absent-minded and there are many stories about that. However, one of the mysteries was his 'selective memory' and it therefore made it difficult to know how much he remembered of the last conversation you had with him. As a fresh research student I had the honour of being seated next to him at the matriculation dinner in Caius College, where he was then Master. The conversation rapidly moved to the topic of my research, the electronic properties of graphite; all I knew was that graphite was black and behaved rather like mica. Being young and brash, I was pre-pared to talk and speculate about any of its properties. He was obviously interested and asked if I would call on him the next day at 10 A.M. to continue the discussions away from the noise of clanging glasses and silver. I thought I had got myself really trapped this time and spent the night frantically reading up everything I could lay my hands on about graphite. The hour came in the morning when I gingerly knocked on his door. I entered while he was reading some paper and did not raise his head. The few seconds I waited seemed like hours, and when he eventually got round to looking up, he said 'Who are you?' The meeting thus terminated almost as soon as it started, as he seemed to show absolutely no interest either in me or graphite. Secretly I was pleased and greatly relieved to have been spared further embarrassment of having to show off my ignorance.

We did not have any more scientific exchanges in the next four years as I was working in the Physics and Chemistry of Solids (PCS) Laboratory, which was geographically remote from the Austin Wing, where he headed the depart-ment. However, one day in 1967 I received a surprise telephone call from him asking, 'You are the man working on graphite, aren't you? I'd like you to come to my office to talk about carbon bonding.' A different kind of problem now got hold of me as I had by then changed fields to work in II–VI com-pounds and layer-type transition metal dichalcogenides. Nevill had cast me in the role of an expert on graphite and threw me in among Phil Anderson and Volker Heine. I cannot remember much of the outcome of the discussion. Perhaps I was just glad to have emerged alive from his room.

After he resigned his Mastership of Caius in 1966, he devoted much of his time to the topic of metal–insulator transitions which led to disordered solids

Nevill with Professor Yao Liang at the IRC in Superconductivity during the 1990s.

and also to establishing amorphous materials as a new discipline. There are others in this book who are better qualified to tell those stories. His resignation from the Mastership of the College at that time caused quite a stir in Cambridge as this kind of thing does not happen very often. I think Nevill did not have the patience to deal with college people who treated college life very personally, and a small group of people like that could greatly influence affairs in the college. In any case, dropping his magisterial responsibility in the college did physics a great favour by his not only having created a new topic in solid-state physics, but also winning a Nobel prize.

His retirement from the headship of the Cavendish Laboratory coincided roughly with the move of the laboratory to the West Cambridge Site. By this time Nevill was ready to tackle a new subject; he picked low-dimensional physics in the form of artificially created inversion layer in semiconductors, and in the rich field of layered transition-metal sulphides and selenides. He was offered an office by Abe Yoffe in PCS next to my room, and it was a pleasure for many years to have Nevill within reach for discussions. He had an uncanny ability to identify the key questions of a very complex subject. In the case of transition-metal layer compounds, this was demonstrated by his pre-occupation with the question, 'How can tantalum disulphide (TaS_2) be a semi-conductor and at the same time display Pauli paramagnetism?' He frequently repeated this puzzle in seminars and urged Abe Yoffe to set up a high-pressure facility to measure the transport properties of tantalum disulphide. The timing is important here. He was raising this question ten years before the observation of charge density waves in tantalum disulphide, which we now know are responsible for its many unusual properties, and which for some time in the late 1970s and early 1980s was an important field in the study of condensed matter.

The discovery of high-temperature superconductors in 1987 was an important turning-point in his scientific life, and this gave him much pleasure and satisfaction in his twilight years. But make no mistake, Nevill approached this exciting field with a fertile and creative mind, and with the zest of a young man. We had many opportunities to co-author papers on the subject of high-temperature superconductivity, but never did so. I believe, however, that I did introduce him to it, by initially trying to interest him in the metal–insulator aspect of the transition, and discussing with him my first paper on the subject. Nevill, however, had greater ambitions beyond metal–insulator transitions; he wanted to solve the mystery of high-temperature superconductivity and provide a full explanation for its microscopic mechanism, as he had done for so many other topics. Whether he was right in his interpretations and model still remains to be proven, but his impact on the subject and his contribution to highlight the importance of the charged Bose liquid is beyond question. How many scientists could claim so many significant papers and books written for a new and major field well into their nineties?

The discovery of high-temperature superconductivity also brought other exciting developments to Cambridge. The Interdisciplinary Research Centre (IRC) was established in 1988, and taking the opportunity of space reorganisation in the Cavendish, Nevill asked me if he could have a room among the superconducting researchers at the IRC – he wanted to feel the pulse of high-temperature superconductivity research by being at the heart of where it was all happening. Here was a man who turned his back on a comfortable office in a building named after himself to take up a tiny cubicle in temporary accommodation created for the IRC in Superconductivity, where the room temperature could rise to the unbearably high thirties (Celsius) in the summer. (Coincidentally another Emeritus Cavendish Professor, Sir Sam Edwards, now occupies the same cubicle.) He was a wonderful gain to the IRC and I was pleased to have played a small part in bringing this about, just like Abe Yoffe fifteen years earlier. Our accommodation improved enormously when we moved to our present building. The IRC received generous funding from its research council, and support from the Leverhulme Trust facilitated Nevill's last scientific partnership with high-powered Russian theoretician Sasha Alexandrov. This partnership was highly successful as they had much to gain from each other, at a scientific as well as a personal level. He continued to work single-mindedly on high-temperature superconductivity to his last days. It would have been much more gratifying had he lived to see a more general acceptance of his theory of high-temperature superconductivity; certainly he would have liked more of his old colleagues converted to his view. Nevertheless, this view is very much alive and continues to gain momentum.

Knowing Nevill and sharing the many precious moments with him, both inside and outside the sphere of science, has been for me a uniquely privileged and enriching experience.

PROFESSOR W. Y. LIANG was born on 23 September 1940 in Indonesia, where he lived until coming to the United Kingdom in 1958. He read physics at Imperial College, London, and carried out his PhD research under Abe Yoffe at the Cavendish Laboratory, where he is now Professor of High-Temperature Superconductivity. Since 1989 he has been Director of the Interdisciplinary Research Centre in Superconductivity at Cambridge.

High-Temperature Superconductivity

Sasha Alexandrov

Is there an Explanation?

It was an eventful year, 1986. The events that took place had a tremendous impact on the lives of many people, perhaps even nations. Among them was the discovery of new superconductors by Swiss physicists J. G. Bednorz and K. A. Müller, a discovery that will probably transform our lives well beyond the twentieth century.

Superconductivity is an 'old' phenomenon discovered by Kamerlingh Onnes in Leiden in 1911. He observed a totally unexpected disappearance of the resistivity of mercury below the 'transition' temperature T_c of 4 K ($-269°C$). Subsequent work revealed that many metals and alloys displayed similar properties, the transition temperature of Nb_3Ge at 23 K being one of the highest. Despite numerous efforts, no adequate explanation of this phenomenon was suggested, until the work by Bardeen, Cooper and Schrieffer (1957). By that time a superfluidity of neutral liquid helium He^4 had been discovered below a T_c of 2.2 K. Ever since the discovery of quantum mechanics it has been known that the helium atom is a boson, and that the atoms of liquid helium should obey Bose–Einstein statistics. The theory by Bogoliubov (1947) explained how these statistics led to a frictionless flow of the liquid. Liquid helium could now be described by a two-fluid model: below the transition temperature T_c it could be considered as a mixture of a superfluid, without entropy, and a normal fluid. A crucial demonstration that superfluidity was linked to Bose particles and Bose–Einstein condensation came after experiments on liquid He^3, whose atoms were fermions. He^3 failed to show the characteristic λ transition within a reasonably wide temperature interval around the critical temperature for the onset of superfluidity in He^4. The idea of bosonization of electrons due to their pairing was introduced by Ogg (1946). It became very attractive as an explanation for superconductivity after Schafroth (1955) showed that a gas of charged bosons with the charge $2e$ would indeed show an expulsion of the magnetic flux (the Meissner–Ochsenfeld effect), which was a characteristic feature of superconductivity. However, this phenomenological Bose-gas picture was condemned and later on practically forgotten, because it failed to account quantitatively for the critical parameters of classical (low-temperature) superconductors. Using the Bose–Einstein condensation temperature for T_c one obtained an utterly meaningless result, $T_c = 10000$ K, with the density of electron pairs at approximately 10^{22} cm^{-3}, and

Coffee break with Sasha Alexandrov at the IRC in Superconductivity during the 1990s.

with the effective mass of each boson at twice the electron mass, $m^{**} = 2m_e$. The inapplicability of the 'real-space pair' picture resulted from a very large size of pairs in classical superconductors.

The Bardeen, Cooper and Schrieffer (BCS) theory was based on a 1950 demonstration by Fröhlich that electrons in states near the Fermi surface of a metal could attract each other weakly on account of their interaction with phonons. Cooper showed that electron pairs could thereby be formed with a diameter much greater than the lattice parameter, and stable only due to their quantum interaction with other pairs. The final theory showed that in a small volume of momentum space round the Fermi surface, the electrons would form pairs, or bosons, all having the same energy. These 'momentum-space pairs' would strongly overlap in real space in contrast with the model of non-overlapping real-space pairs. Highly successful for metals and alloys with low T_c, the BCS theory led some theorists to a wrong conclusion that, by the nature of things, there could be no high-temperature superconductivity with T_c above 30 K, which implied that Nb_3Ge already had the highest T_c. Although the Ogg–Schafroth phenomenology led to unrealistically high T_c values, the BCS theory left no hope for discovering new materials which could be super-conductors at room temperature or at least at liquid nitrogen temperatures (about 70 K).

However, well before the experimental discovery of high-T_c superconduc-tors, it was realized (Alexandrov and Ranninger 1981a, b) that the descriptions were actually two opposite extremes of the same problem of electron–phonon interaction. By extending the weak-coupling BCS theory towards the strong interaction, a Bose gas of small bipolarons was inevitably predicted. Subsequent studies (Alexandrov 1983) led us to the conclusion that high temperature superconductivity should exist in the crossover region of the interaction from the polaronic BCS-like state to what is now called the bipolaronic superconductivity of real-space bosons. Our estimate showed that it was possible to achieve a critical temperature as high as 100 K in this region of electron–phonon interaction (Alexandrov and Kabanov 1986). One can imagine my own and my co-workers' excitement when Müller and Bednorz wrote in their original papers (Bednorz and Müller 1986), and later on stated in their 1988 Nobel prize lecture, that in their search for high-T_c superconductivity they were stimulated and guided by the polaron model. Their expectation was that if 'an electron and a surrounding lattice distortion with a high effective mass can travel through the lattice as a whole, and a strong electron–phonon coupling exists' the perovskite insulator could be turned into a high-temperature superconductor. Many, but not all, incorporated the concept of a bipolaron in their theoretical and experimental studies of new superconductors, which now include low-dimensional mercury cuprates with the highest T_c of 160 K (under pressure) and three-dimensional doped fullerenes ($M_x C_{60}$) with T_c of 40 K as well. Among those who considered the model realistic were prominent Russian physicists V. L. Ginsburg, A. I. Larkin, D. Khmelnitskii and E. I. Rashba.

On the other hand, many others remained sceptical. The problem was not with the theory itself, which was as good as any correct theory, but by accepting our theory one had to recognize that the BCS approach was not universal. Moreover, it meant the 'old' generally accepted extension of the BCS theory to the strong-coupling limit was wrong. Soon after the discovery, one of those sceptics told me in a friendly conversation that it would be quite improbable if the new materials were not Fermi liquids to which the BCS theory could apply. However, impressed by the rapid increase of the T_c value, those who argued that high-temperature superconductivity was impossible decided to turn their minds toward the non-phononic mechanisms of pairing. Many young workers began to be involved in the search for the BCS-like superconducting ground state of low-dimensional fermions with a pure repulsive (Coulomb) interaction. The deteriorating situation in Russian science, the growing competition for an explanation of the high-T_c phenomenon and my own heavy administrative workload (at that time I was Research Vice-Rector of the Moscow Engineering Physics Institute, MEPI) encouraged little optimism.

Precisely at that time, Yurij Kopaev of the Lebedev Physics Institute in Moscow told me about Mott's letter in *Nature* (Mott 1987). It was like a ray of hope for me and a few others. For many of us, Sir Nevill Mott was a 'godfather' of condensed matter physics. I knew his excellent book written with Ted Davis, *Electronic Processes in Non-Crystalline Materials* (Mott and

Davis 1979). The elegance with which Sir Nevill derived his famous law for temperature dependence of the conductivity of disordered solids, even inspired a joke about 'his direct communication with God.' In his letter to *Nature* entitled 'Is there an explanation?' he wrote, 'The short answer is that there are many, perhaps as many as there are theorists active in the field.' However, he went on, 'Many, but not all, use the concept of the bipolaron.... The questions arise, is a Bose condensation of bipolarons possible, and would it be superconducting? Alexandrov *et al.* considered the possibility in some detail in 1986, and concluded that the answer to both questions was yes.' Only then did I realize that we were not alone and there was a strong support emanating from one of the most creative physicists of our century.

Paris in June and Cambridge in December

Mott's letter in *Nature* gave a new impetus to our efforts at the MEPI towards a complete theory of newly discovered superconductors. However, I clearly realized that to compete in the field I had to choose between basic research and my administrative and academic career. The number of 'superconducting' papers both experimental and theoretical was growing exponentially from year to year. Many of the active physicists working in the field were leaving Russia for Western Europe and the United States. However, each time I travelled abroad I had to overcome the barriers, which consumed a lot of time and energy, leaving less and less for my research. The solution arrived with a letter from Herbert Capellmann inviting me for a research position at the Aachen University, Germany, which I received in November 1989. Herbert's invitation and the understanding of my 'dilemma' by the MEPI Rector, made it possible for me, my wife Elena and our son Maxim to cross three borders, driving our tiny Lada in March 1990 to meet with Herbert and our new life in Germany. It was then that I realized there were several groups in the West which were already involved in studies of the bipolaron model of high T_c.

In particular, Dipak Ray of Paris University pointed out in his *Phil. Mag. Letter* (1987) that, by good fortune, the width of the electronic band in cuprates was just right for bipolaronic superconductivity, i.e. not too wide and not too narrow. He invited me to spend a month in Paris, which turned out to be a crucial point of my scientific career. In fact, Dipak was responsible for bringing me and Mott together in June 1990. I remember the sunny day when I met Sir Nevill in Dipak's office. Mott put his hand on my shoulder and asked me to come closer to the window. He explained that because of his age (he was 84) he could not see my face clearly in the dark. I was really impressed by the firm grip of his hand and the youthful twinkle in his eye. In front of me stood a tall, handsome man radiating health and confidence. He was surprised as well. 'When I was reading your papers I didn't think you would be such a young lad,' he told me. At that time I was almost 44. We spent this day discussing our approaches to the high-T_c problem. We agreed immediately on what united us, a Bose-gas picture of high-T_c cuprates irrespective of the way fermions were coupled. I believed and continue to believe that the electron-phonon interaction is responsible for the pair formation in cuprates. Impressed by the first experiments which had not shown any isotope effect, Sir Nevill thought that spin fluctuations of the antiferromagnetic background played the main role in bipolaron formation. In one of his papers Mott (1990) wrote:

> As I have already stated, before the discovery of the high-temperature materials, Alexandrov *et al.* suggested that superconductors might exist in which electron pairs were the current carriers, that they were bosons and formed a condensed phase below T_c, that they would act as superconductors and that the carriers above T_c would be bosons, or if these dissociate, a mixture of bosons and fermions. ...
> In this article I adopt this model and advocate a particular mechanism for the formation of pairs, namely that of the existence of spin polarons.

Already at that time we agreed that the mechanism of bipolaron formation was of secondary importance. Whereas what was and continues to be of primary importance is the breakdown of the Fermi liquid and the BCS theory because of the existence of pairs above T_c. Subsequent experiments showed a finite and sometimes large isotope effect both on T_c (Franck 1994) and on the Néel temperature (Zhao, Singh and Morris 1994). While writing our last book together (Alexandrov and Mott 1995), Sir Nevill humorously waived the argument in favour of the electron–phonon interaction:

> Theories stand or fall by their ability to explain, or better to predict, experimental facts. The behaviour of the new superconductors, whether ceramics or single crystals, has proved to be very sensitive to the technique of preparation, and I am aware that the experimental results quoted may appear somewhat selective. The late John Hubbard, a distinguished theorist, used to say 'you can prove anything by experiment'.

Nevertheless, Mott finally agreed that both coupling mechanisms could operate at the same energy scale, so in our book we assumed the point of view that carriers in doped Mott–Hubbard insulators were small bipolarons surrounded by both the spin and the lattice polarized regions. At the end of our stay in Paris, to my greatest delight, I was invited to give a talk at Cambridge.

The first visit to Cambridge in December 1990 was absolutely fantastic. After a week in Turin, Italy, where I presented a paper at a conference, my family and I went up to Leiden, the Netherlands, at the invitation of Professor de Jongh, who had advocated bipolarons in cuprates in several publications and asked me to give a talk at a seminar at the Kamerlingh Onnes Laboratory. The day after this seminar, having overestimated the capacity of roads other than the German autobahns, we were fifteen minutes late for the ferry in Ostende to take us across the Channel. The next ferry we took was five hours late itself, due to a failure in its lifting equipment. So it seemed all odds were against arriving on time in Cambridge. Here I must say I always trusted my wife's ability to grapple with problems. In fact, having docked at Felixstowe almost five hours behind schedule and with only an hour and a half before we were due at the seminar, Elena driving on 'the wrong side of the road' managed to take us right to the Cavendish. We arrived with just fifteen minutes to spare; still waiting for me were Sir Nevill and many people from the IRC in Superconductivity. For the three of us, those two unusually cold and snowy days wrapped in the generous warmth of Gonville and Caius College seemed like stepping into a fairyland. Sir Nevill, who drove himself at that time, was enthusiastic about showing us around the beautiful city as well as his famous college. During the two days of invigorating discussion of bipolarons and high T_c, I was overwhelmed by the vast creative potential of this scientist, potential far from being exhausted. We both realized that it was absolutely essential to put our minds together to develop a complete theory of new materials.

Last Four Years

In spring 1992 a letter came from Yao Liang, Director of the IRC, telling me that his and Sir Nevill's efforts to arrange for our collaboration turned out to be surprisingly successful. The Royal Society established a new Mott Bye-fellowship, and I was invited for the position paid first by the Leverhulme Trust and subsequently by the Newton Trust and Gonville and Caius College. This time, the Seacat crossing from Calais to Dover took us only forty minutes and the remaining four years with Sir Nevill passed as fast as this voyage. Genius he was, but a hardworking genius. In less than four years we had published two books (Alexandrov and Mott 1994, 1995) and coauthored nine papers, five of which appeared in *Physical Review Letters*. At the same time, Sir Nevill produced about the same number of papers on his own. He would usually show up at the IRC twice a week, driving all the way from Aspley Guise – quite a distance – and staying well into the evening. Only in the last two years did he stop being his own driver, but he still kept to his weekly schedule. Most of that time we spent in discussion and writing on the blackboard, which was actually white and which he used quite extensively. There were visitors as well, many coming from the next-door Cavendish Laboratory, and others from the wider world, all willing to share their insights into the intricate world of modern physics with one of its active luminaries. It was due to Sir Nevill's visitors that I was able to put faces to so many prominent names. Sir Nevill's standing observation about Cambridge was that its density

313

UNIVERSITY OF CAMBRIDGE
DEPARTMENT OF PHYSICS

From
PROFESSOR SIR NEVILL MOTT, F.R.S.

CAVENDISH LABORATORY
MADINGLEY ROAD
CAMBRIDGE CB3 0HE
Telephone: 0223 - 337733
Telex: 81292

26.1.93

Dear Sasha

On p 225 of my 'Metal- Inst. Transition 1990 (2nd ed) is a figure of the binding energy of two spin polarons, calculated by Su → Chen, Phys Rev B 38, 8071 (1988)

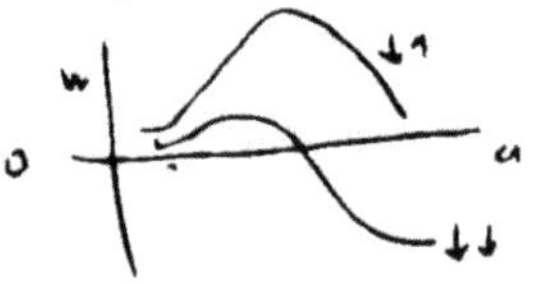

This seems relevant to the Buckel (Zurich) paper, that has kink ρ ... in the p-T

(if the bonding ρ is determined by my spin polaron model, the bipolaron is more spread out, and so smaller effective mass?

Incidentally, if round 200 K the bipolarons become overcrowded (liquid) should not John Harris see the appropriate entropy?

I hope to see you next week

Nevill

Original letter from Nevill to Sasha Alexandrov (1993).

Sasha Alexandrov presents Nevill with a selected volume of his papers; Abe Yoffe is seated next to Nevill.

of Nobel prizewinners per head of population was the highest in the world. He was very proud of Cambridge and of being part of it.

During the rest of the week he would work at home, so practically every week I would receive a letter or two generating new questions and ideas. I admired those letters. Sometimes they would give a complete clear-cut explanation of various properties of cuprates. One of them particularly impressed me. We thought that the gap, which used to be identified as a superconducting gap, was actually the bipolaron binding energy. If so, the gap should be temperature independent, existing well above T_c and its value should not be directly dependent on T_c. The reflectivity measurements of the gap by Schlesinger *et al.* (1991) perfectly agreed with the assumption. However, a possibility remained that the 'normal state gap' did not belong to the charge carriers, but was a gap in the spin excitation spectrum, as suggested by some theorists.

Hence it was crucially important for us to find experimental evidence for the normal state gap in the charge transport. In January 1993 Mott wrote to me, isolating this evidence from the in-place resistivity data of the Zurich group. His original letter is reproduced here.

In just a few sentences he explained the observed kink in the ρ–T plot by a coexistence of singlet and triplet bipolarons. The triplet was more spread out, and therefore had a smaller effective mass. Increasing temperature meant that more and more singlets became triplets, which explained a drop in the resistivity slope with temperature. Moreover, in this short letter he predicted the effect of triplets in the electronic entropy, which led us towards an explanation of the normal state electronic heat capacity of cuprates measured by the IRC group (Alexandrov and Mott 1994).

My collaboration with Mott was not interrupted at all when in summer 1995 I was invited by Professors David Wallace and Kurt Ziebeck to the Chair in Theoretical Physics at Loughborough University. The excellent conditions in my new Physics Department and the fruitful contacts with experimentalists at both Loughborough and Cambridge only boosted our research. During Nevill's last two years we completed our final book, published by World Scientific. The same publisher invited Mott to edit the twelfth volume of the World Scientific Series in Twentieth-Century Physics, a selection of his papers. He finally agreed on the condition that I would help him to choose. *Sir Nevill Mott – 65 Years in Physics* was published in record time, and the first copies were available on Nevill's ninetieth birthday. Inside Mott wrote: 'The new superconductors, such as $La_{2-x}Sr_xCuO_4$ seemed to me doped semiconductors, which I studied since Mott and Twose. Working with A. Alexandrov, we have what we think is a complete model of these materials, where we think the current carriers are bosons (bipolarons).'

Quite often Sir Nevill told me that he believed in science and he accepted religion and the idea of God, but what he could not accept were dogmas, both religious and scientific. Along with David Khmelnitskii, I tried to convince him not to comment on the alternative theories of high-T_c in the press. But all our efforts failed and he published his letter in *Physics World* (Mott 1996). He explained to us that his comment was addressed to the younger generation of scientists starting their careers rather than to our generation. Nevertheless, Mott's letter gave me a strong impetus for an understanding of the unusual temperature dependence of the out-of-plane resistivity, one of the key non-Fermi-liquid features of cuprates. One day, reading Mott's letter again, I realised that only thermally excited polarons could contribute to the c-axis transport at intermediate and high temperatures because they were much lighter in the c-direction than bipolarons. Along the planes they propagated with about the same effective mass as singlets. As a result, we had a mixture of non-degenerate quasi-two-dimensional spinless bosons and thermally excited fermions, capable of propagating along the c-axis. A simple fundamental relation between the anisotropy and the magnetic susceptibility immediately followed from the assumption. Fortunately, at that time my former PhD student

Viktor Kabanov was with us at Cambridge. Less then a month later all calculations and the experimental fit were done, and in June 1996 we submitted our letter (Alexandrov, Kabanov and Mott 1996) to *Physical Review Letters* (PRL); it proved to be the last paper coauthored by Sir Nevill. He planned to report the result at the seminar in Orsay on 3 October 1996, to which both of us were invited by Professors Friedel and Jérome. On Thursday 8 August, upon arriving at the lab, I found a letter from the PRL editor with two positive referee reports. I thought how pleased Sir Nevill would be when he saw these reports; I was expecting him to arrive at the start of a normal working day. Within half an hour, Mott's daughter Alice called to say that Nevill had died suddenly from a heart attack.

On 2 October I left for Paris alone; I was flying to the place where our collaboration had begun.

References

ALEXANDROV, A. S. (1983) *Zh. Fiz. Khim.*, **57**, 273; *Russ. J. Phys. Chem.*, **57**, 167 (1983).

ALEXANDROV, A. S. and KABANOV, V. V. (1986) *Fiz. Tverd. Tela (Leningrad)*, **28**, 1129.

ALEXANDROV, A. S. and MOTT, N. F. (1994) *High Temperature Superconductors and Other Superfluids*, London: Taylor & Francis.

ALEXANDROV, A. S. and MOTT, N. F. (1995) *Polarons and Bipolarons*. Singapore: World Scientific.

ALEXANDROV, A. S. and RANNINGER, J. (1981a) *Phys. Rev. B*, **23**, 1796.

ALEXANDROV, A. S. and RANNINGER, J. (1981b) *Phys. Rev. B*, **24**, 1164.

ALEXANDROV, A. S. KABANOV, V. V. and MOTT, N. F. (1996) *Phys. Rev. Lett.*, **77**, 4796.

BARDEEN, J., COOPER, L. N. and SCHRIEFFER, J. R. (1957) *Phys. Rev.*, **108**, 1175.

BEDNORZ, J. G. and MÜLLER, K. A. (1986) *Z. Phys. B*, **64**, 189.

BOGOLIUBOV, N. N. (1947) *J. Phys. USSR*, **11**, 23.

FRANCK, J. P. (1994) in *Physical Properties of High Temperature Superconductors IV*, ed. D. M. GINSBERG. Singapore: World Scientific, p. 189.

MOTT, N. F. (1987) *Nature*, **327**, 185.

MOTT, N. F. (1990) *Contemp. Phys.*, **31**, 373.

MOTT, N. F. (1996) *Phys. World*, **9**, 16.

MOTT, N. F. and DAVIS, E. A. (1979) *Electronic Processes in Non-Crystalline Materials*, 2nd edn. Oxford: Oxford University Press.

OGG, R. A. Jr (1946) *Phys. Rev.*, **69**, 243.

RAY, D. K. (1987) *Phil. Mag. Lett.*, **55**, 251.

SCHAFROTH, M. R. (1955) *Phys. Rev.*, **100**, 463.

SCHLESINGER, Z. *et al.* (1991) *Physica C*, **185–189**, 57.

ZHAO, G., SINGH, K. K. and MORRIS, D. E. (1994) *Phys. Rev. B*, **50**, 4112.

PROFESSOR A. S. ALEXANDROV was born in Novgorod, Russia, in 1946. He was Professor of Physics, Dean of Theoretical Physics and Vice Rector at the Moscow Engineering Physics Institute (MEPI) before joining Nevill Mott at the Interdisciplinary Research Centre (IRC) in Superconductivity at Cambridge in 1992. He was appointed to the Chair of Theoretical Physics at Loughborough University in 1995.

Religion

My Brother-in-Law

John Horder

This is a chance to think again about Nevill Mott, my brother-in-law, and in a more concentrated way than ever before. I am grateful.

He came into our family when I was seven and he was twenty-two. That gap in age was still there for me through much of the seventy years since then. I felt small beside him, but I am sure that he did not want it that way. He was always kind to me. But the gap was kept open through my abysmal ignorance of physics and mathematics, so what we had in common was not what seemed to be most central in his life. But we did share music from the start (for a time Nevill played the clarinet, well enough to manage the Mozart Quintet) and we did share a love for French Gothic cathedrals (he made a very competent model of one from cardboard and paper). I shall never forget his later invitation to a Dr Caius dinner in the fifties. Four hours of wonderful food and music must have made both of us feel more at ease and more expansive than usual. That was when he first revealed to me that his name might be proposed for the Mastership of the college and how pleased he was.

In the last thirty years it was different. The change was gradual, helped no doubt by my late discovery and admiration for the discipline of scientific thinking (although medicine has been my profession, my education until the age of twenty, like Ruth's, was entirely in the humanities, without a single hour of any science). But the main reason was that we now shared an interest in religion, although we came to this from opposite directions. Nevill described his upbringing as atheist. Mine had been intensely Christian – Congregationalist at home, Anglo-Catholic at boarding-school. In principle he was moving in and I was moving out.

He made a habit of sending me typescripts of what he wrote about religion or what he said in the pulpits of Aspley Guise and Cambridge between 1983 and 1994. I have been re-reading all of them to see if I can understand any more clearly what drove him in this unexpected direction and whether his thinking can help others who are concerned to pursue both reason and Christian belief without abandoning the essence of one or the other.

What drove him? First, I think, a sense of something missing in his life. He expressed it most easily as a need for continuity with the past, with unchanging traditions and with beauty: 'the hauntingly beautiful liturgy and hymns of the Church of England.' Or again: 'I can share in the worship conscious of a sense of history.' He asked: 'Do you not feel that for us too the Christian churches that we inherit are infinitely precious? . . . I remember, as a young man, pushing my bicycle to the top of a hill in Norfolk and seeing seven church towers in the surrounding country. . . . Could all Christian doctrine be false when it had inspired an age that has poured its meagre resources into the creation of so much beauty, to the Glory of God? The great cathedrals tell you the same.'

Secondly, certain people influenced him: Mervyn Stockwood, who moved as he did, from Bristol to Cambridge, and later Hans Küng in Tübingen. Others influenced him too, including my sister, Ruth, although her influence was perhaps the least visible. He wanted to find that Christian doctrine was not false and these were people who held his love and respect, but had visions which were remote from those of the physicist. They might supply what had been missing before.

Thirdly, Christian beliefs, despite their claim to the authority of revealed truth, offer a fascinating intellectual challenge, because of the difficulty of aligning their content with knowledge acquired during the last two thousand years.

In summary, Nevill felt the need for an historical tradition wider than that of physics, 'which for me has little to do with the search for God.' He found one which is inevitably full of problems for any thinking person and used his exceptional intellect to help both himself and others who find it impossible to put them out of mind. How far did he succeed in this?

Most of his thinking is contained in his contribution to *Can Scientists Believe?* In this book, which he edited, Nevill declared his need for belief in God, but the impossibility of believing in miracles which imply that God is able and willing, at times, to overturn the laws of nature:

> I believe that the laws of physics and chemistry are not broken; water is not turned into wine and a body is not removed miraculously from a tomb. I must believe, if God is omnipotent, that he could do these things if he wanted to, but I cannot worship or respect a God who would want to. . . . The idea is repellent that God should set aside his laws on special occasions, as, for instance, arranging a virgin birth for Jesus, and yet allowing the Black Death, the Holocaust, and other horrors to occur. My assumption is that God . . . works within natural law.

Clearly there was no question of literal acceptance of the Bible or of every traditional Christian belief: 'Uniform belief is impossible for modern, educated people, unless they close their minds to the meaning of words.' He could not have sought membership of the Roman Catholic Church. Indeed he rejected several articles of the Creed of the Church of England, to which he was admitted on the simple statement: 'I believe in God and understand him through Jesus.' At the end of his contribution he wrote:

> I most strongly believe that all individuals should be free to believe only those doctrines which help them in their approach to God. . . . I

want to belong to a church which realises that there are many approaches to God and can extend wisely the limits of belief, but at the same time preaches a God external to ourselves who, through the teaching of Jesus and his followers through the centuries, opens to us our relation to Him.

For Nevill, God is not only external to ourselves but is described, according to tradition, as a person who can be addressed. But is He also the creator of the universe, the laws of nature and the process of evolution?

> Is there a relationship between the God to whom we can pray and the 'God' who created the universe, the God of power? Some people say that the omnipotent God has voluntarily relinquished his power and this may be a helpful concept. But I would rather say it is a mystery, something we do not understand, have no way of understanding. . . . I do not believe, or wish to believe that God's hand is to be seen in the steps of the evolutionary process; were it so, I would have to ask why He allowed plagues and earthquakes, and fail to see in Him the loving father of Jesus Christ. . . . In the creation I see little of the nature of the loving God; the Old Testament's 'God saw that it was good' rings hollow to me. Natural is neutral, as emphasised so strongly by Albert Schweitzer. But we can see the beauty of mountains, forests and cathedrals as part of the miracle of human consciousness, given by God.

So it is in human consciousness, 'this mystery existing somehow alongside natural law,' that Nevill sought the relationship between God and man and the place where God's action can be traced:

> To give meaning to consciousness, a belief in God who is outside us is necessary to me. Without Him life can seem a tale told by an idiot. I believe in God because I wish to do so, to give meaning to human life.

He had no difficulty in arguing that consciousness, awareness of self and human free will cannot be explained or understood through the laws of physics and chemistry. Quoting Brian Pippard:

> It is impossible that a theoretical physicist, even given unlimited access to the most powerful computer imaginable, should deduce from the laws of physics that a certain complex structure, namely himself, should be aware of its own existence.

There is nothing in the laws of physics and chemistry (or of psychology) which prevents a belief in human free will.

God's only power is His influence on men when they ask Him for it through love, Intercessionary prayer is something which brings a sick or unhappy person before us, rather than something which will change his or her fate. And Jesus?

Here was a man who believed in the stories of the Old Testament, that the Parousia (the second coming) would come soon, and who taught a sublime other-worldly way of living. Perhaps some Christians see him more clearly than they see God the Father and pray to him first. I cannot do this. The world he lived in is too remote. And can I say that *uniquely* there is more of God in Jesus than in any other man? And was the doctrine of Jesus a *unique* proclamation of God? If so, why did he address only the Graeco-Roman world?

Nevill goes on to describe his difficulties with the doctrine of redemption from sin through the crucifixion of Jesus, the doctrine of eternal life and the introduction of the principles of the Sermon on the Mount into the way we live.

I have tried to reflect Nevill's thinking as accurately as I can through quotations from what he published. What seems clear is that he neither evaded difficult problems nor abandoned the form of rational thinking which belongs to the sciences. His honesty, concern and intellectual courage are apparent. But there is disparity between his expressed beliefs and the short statement of Christian doctrine which he includes in the introduction to *Can Scientists Believe?* No one would now wish to question whether or not he retained the essence of Christianity, although in the book he does question this himself.

His clarity has helped me to understand a little more clearly why I cannot pursue the same path. I do not feel the same need nor share the most central of his beliefs, on which the rest depend. But I have shared the problems with him and shall always feel grateful that this brought me closer to him.

DR J. HORDER, CBE, is Ruth Mott's brother. He was born in 1919 and served as President of the Royal College of General Practitioners and Vice-President of the Royal Society of Medicine before his retirement. He is Honorary Fellow of Green College, Oxford, and of Queen Mary and Westfield College, London.

A Scientist who Came to Believe

Christa Pongratz-Lippitt

My parents got to know the Motts in Bristol before the war. I do not even know whether my earliest memories of Sir Nevill are my own, or whether I merely think I remember him from my mother's wartime stories. But it was in Bristol and I suppose I must have been about eight years old. War had been declared, as I distinctly remember that I had dropped some eggs on the way home from the grocer's that morning, and had been told to be more careful in future as there was now a war on and eggs would soon be rationed. Later that day, I was told to say hello to a very, very tall man – one of the tallest men I had ever seen up to then. He had a frightful cold but was smiling, and I was duly impressed but not a little puzzled on being told, whether then or later I cannot say, that he was 'jolly good at maths but a simply rotten speller.' This obviously intrigued me the more because I found spelling easy but was simply rotten at sums. But the fact that he was smiling in spite of a ghastly cold became one of my mother's wartime stories, which she loved to tell until she was over 90. 'He should have been in bed, you know, with a terrible cold like that,' she would say when talking about Bristol in the early days of the war, 'but there was a war on.' And she never failed to stress how stoical Sir Nevill had been. And so, since that first encounter with Sir Nevill at the age of 8, I knew that however bad a cold one had, one must keep smiling.

I don't think I can have seen the Motts again until after the war. I had got a place at Lady Margaret Hall (LMH), Oxford, to read modern languages and wanted to read Russian, but LMH preferred 'just normal languages' in those days, as then I would have a moral tutor in college. They were also hesitant to let me start a language from scratch, although it was allowed. My mother suggested I go and talk things over with Sir Nevill, who was still at Bristol at the time (1949). He could not have been more encouraging. Over tea and something delicious like hot scones, he supported me one hundred per cent. Rationing was still very much on people's minds, so whatever it was that we had for tea must have been a great treat, otherwise I would not to this day remember how delicious it was. He offered to write to LMH immediately, saying maybe that would help. He spoke enthusiastically about Europe and European union, and how important it was to learn languages, above all Russian. This would enable one not only to read Russian literature and Russian publications, but one could then talk to the few Russians who were allowed out to attend conferences, and so help to break down barriers. Sir Nevill himself spoke German well and it was on that visit that he told me how he had taught himself German in his undergraduate days at Cambridge during the twenties, in order to be able to read the papers of Erwin Schrödinger, the Austrian physicist who was awarded the Nobel prize for physics in 1933. Whether it was his letter to LMH that did the trick, I do not know, but they let me read Russian after all.

Years passed. Occasionally my mother would mention that she had heard from the Motts, but it was not until after her death in 1992 that Sir Nevill and I began to correspond. He, meanwhile, had become a Christian and was now a member of the Church of England, and I, a born and practising Catholic, was Vienna correspondent of the international weekly the *Tablet*. 'It is an excellent journal,' he wrote in one of his first letters, 'and I have several quite intimate friends who are Catholics. I do understand the need for certainty which the Roman community satisfies, but I am far from being a Catholic. Locally here, however, we are all – Anglicans, RCs and non-conformists – very friendly.' We began to discuss matters of faith in our letters, and once, when we were discussing the Sermon on the Mount, he wrote:

I remember reading somewhere that it was addressed to the apostles in an era when the parousia [the second coming of Christ to denote the end of the world, which the early Christians thought was imminent] was expected. It seems to me that the actual decisions one has to make are between grey and grey. Would someone who really tried to follow the Sermon on the Mount make a career in science?

Sir Nevill had met Pope John Paul II in 1987 at a meeting of Nobel prizewinners in Rome arranged by Nova Spes, an organisation founded by Austrian Cardinal Franz Koenig for promoting the global development of man and society. 'I think Nova Spes' aims are excellent,' he wrote, 'and I very much enjoyed meeting its founder, your Cardinal Koenig.' They had sat next to one another at lunch on one occasion in Rome, he said, and had found a great deal to talk about. 'On the question of authority, as you can well imagine, we did not see eye to eye,' Sir Nevill wrote. Papal primacy was, naturally enough to anyone who knew him, something he felt he could not come to terms with, but that did not mean the question did not continue to interest him. He immediately asked me to send him an article by Cardinal Koenig on whether papal primacy was a source of unity or disunity, and later wrote back to say he had found it fascinating.

In November 1992 I got a long letter which began, 'Yesterday our synod (just) passed the measure for ordaining women priests – to my great relief. I could hardly have stayed in a church that voted otherwise.'

Sir Nevill greatly admired the Swiss critical Catholic theologian Hans Küng, whom he first met in 1977 on the quincentenary celebrations of the University of Tübingen. Sir Nevill sat next to Küng at lunch and they later corresponded. When Sir Nevill was reading Küng's thousand-page history of Christianity, he wrote, 'I am fascinated by the history of the papacy, quite unknown to me until now.' Sir Nevill's contribution to a festive anthology for Küng's sixty-fifth birthday was given special praise in the review published in the *Tablet* in March 1993.

I visited Sir Nevill twice in recent years and both times he was there on the station platform to meet me. The first time, at Bletchley station in 1994, I wondered if I would recognise him, as it had after all been forty-five years or more since I had seen him last. And yet I spotted him immediately and like to think it was not because I had seen many photos of him but because I 'recognised' that very, very tall man with the gentle smile, who was so good at maths and such a rotten speller, and whom I had thought heroic at the age of 8 for taking no notice of a bad cold!

I last saw him in September 1995, when he and his sister, Joan Fitch, with whom he was staying in Cambridge, gave me an excellent lunch at a lovely restaurant whose name I cannot recall. We then returned to his sister's house to share a cosy tea with her cat. Sir Nevill was much frailer physically than the year before, I thought, but just as keen to hear all the news from Austria and to tell me all about his Companion of Honour celebrations.

The last letter I had was at the end of October 1995, when he sent me an article he had written on the soul and the brain. He said he had had a month of celebrations on the occasion of his ninetieth birthday and now intended to get back to his normal routine. 'I want to keep up the regime of coming into Cambridge once a week, so as to see Ruth,' he wrote, 'I have had a lot of flattery. How I wish Ruth had been able to enjoy it all with me!'

There are perhaps two things I shall remember especially about him – his gentle smile and his deep devotion to his wife Ruth.

MRS C. PONGRATZ-LIPPITT was born in 1931. Her parents were Austrian and emigrated to England before the Second World War. She went to Dee House Convent in Chester then read modern languages at Lady Margaret Hall, Oxford. After working at the War Office for three years, she married an Austrian and has lived in Austria for over forty years. She has been Vienna correspondent of the Tablet *since 1989.*

Science and Religion

Hugh Montefiore

I came to know Nevill Mott in Cambridge when he became Cavendish Professor, and rejoined his old college of Gonville and Caius, where I was then Dean. He was always approachable as a Fellow and willing to talk to non-scientists like myself. He seemed to me to carry very lightly his deep scientific learning, and he was always gentle in correcting foolish statements by those who did not share such knowledge. He was interested in subjects outside his own speciality; for example, education, university policy, nuclear disarmament. I came greatly to respect him, and when the Mastership became vacant, I suggested to him that he should stand for election. I know that I was not the only Fellow to do this. At any rate, he agreed to stand, and was duly elected. As he admits in his autobiography, this was not the happiest period of his life, but I am still firmly of the opinion that he was the best candidate in the field.

The Fellows of Gonville and Caius had been a troublesome lot even in the time of John Caius himself; he is said to have put some of them in the stocks. By Mott's time, of course, the situation was quite different; but the younger Fellows were both disputatious and democratic. In such a case, a Master was faced with the same kind of difficulties that a Bishop may have when faced with a synodically governed Church. His predecessor at Caius, Sir James Chadwick, another eminent scientist, had also been persuaded to become Master; but the situation so got him down that he resigned.

Mott saw out his Mastership, but although he liked entertaining undergraduates (and clergy from college livings) he was never really happy in the lodge. I think that his main trouble was that he did not always make it crystal clear what he really wanted, and then was upset when the College Council made some decisions which he deplored. However, in his later years, after he had been awarded the Nobel prize, all past difficulties appear to have been forgotten and he seemed greatly to enjoy frequenting the college of which he was then a life Fellow. He kept a room there, even while living in Aspley Guise.

It was through the college that I got to know Nevill and Ruth, his wife. At the time, I was on the College Council (which ran the college) and I was one of the college tutors (who stood in *loco parentis* to the undergraduates, then under the age of consent). Of course, it was in this connection that I met Nevill, but as Dean of the College I also had quite a bit of contact with him. For a person of such a sharp intellect, he had a very relaxed and even laconic way of speaking. I found him unfailingly courteous, and I enjoyed my relationship with him.

I am incompetent to write anything about Nevill's outstanding scientific career, or his management of the Cavendish, which he undertook in addition to his duties as Master. But I can say something about his religious views, which developed greatly while he was in Cambridge and in the latter days of his retirement became matters of major interest to him. I am helped here by the texts of some sermons that he gave and some memoranda he wrote, which have kindly been made available to me.

Nevill's parents did not attend church, so Nevill lacked a religious upbringing, and he was neither confirmed nor baptised. Until he was about 50 he seldom darkened a church porch. He first began to think seriously about religion as a result of an invitation given to him (along with others such as Fred Hoyle) to participate in a course of lectures on science and religion in the University Church of Great St Mary's. Its vicar was Mervyn Stockwood, whom Mott had got to know when he was Professor of Physics in Bristol and Stockwood was a leading figure in that city. Nevill read widely in preparation for the lecture, which meant a lot to him because he reproduced it in his autobiography.

As Master, he used to attend the Sunday evening chapel service in Caius, and later, after his retirement, he began to accompany his wife, Ruth (who was a convinced Christian), to Little St Mary's Church in Cambridge, where he was permitted to take communion. He wrote, 'I come to religion because of my wife, I liked in many ways what I found there and I have been trying for the last twenty years to give meaning to it.'

He decided in his retirement that he would like to be baptised and confirmed, and he wrote out a statement of what he believed:

I believe in God, who can respond to prayers, to whom we can give trust and without whom life on this earth would be without meaning (a tale told by an idiot).

I believe that God has revealed himself to us in many ways and through many men and women, and that for us here in the West the clearest revelation is through Jesus and those that have followed him.

Believing that we should worship God in the company of others, and that here in our beloved England we can best do that within the Anglican church, I ask that the Church admits me through the sacrament of baptism.

My successor as Dean of the College and the Bishop of Ely both judged that this was sufficient, and Nevill was duly baptised and confirmed. What was Nevill's view of God?

Is he something outside us? Yes. . . . Is He an intelligence? Yes. . . . A person (?) We must think of him as such, though I read somewhere that mystics seeking communion with him go through the Dark Night of the Soul as they realise he is not. . . . Perhaps this extract from *The Oxford Book of Prayers* expresses what I want to say: 'God is what thought cannot better; God is whom thought cannot reach, God no thinking can even conceive. Without God man can have no being, no reason, no good desire, naught. Thou, O God, art what thou art, transcending all.'

Nevill believed strongly that we should not change the traditional creeds and worship of the Church:

When we take part in a service in a Christian church we can and do place ourselves in sympathetic communion with the thinkers of the past, and can contemplate the great mysteries through their words as well as our own. So surely we learn from the wisdom of the past, even if – perhaps – we don't believe it all.

It might be thought that, as a scientist, he would have found God in the mysteries of science. Not so.

To me the creation, and the nature of an almighty God, is an insoluble mystery, and by thinking of this, I can learn nothing of the God with whom man can interact, that will help our spiritual life or lead us to right decisions. ... The heart of the Christian religion – I would maintain – is the understanding of the relation of God to man through what we know of the teaching of Jesus, and of all writers in our tradition. Only by meditating on this, I believe, can our religion serve us in helping to decide to which issues, political or otherwise, we can devote our efforts, and give us faith that our work is worthwhile. We come to church or chapel to affirm that faith.

He felt that science gave him a very limited insight into the nature of things, 'and therefore I must have religion. At the same time I seek for a religion that does not depend on miracles and could not be overturned by any historical or scientific discovery.' For Nevill, religion has little to do with historical facts. More than once he quoted Hans Küng: 'Only a person who attaches his faith to historical details will be upset by historical criticisms.'

He was by no means a conventional or orthodox Christian. He was happy to regard doctrines as myths, although he respected them as wisdom expressed in the language and thought forms of an earlier age. He defined his concept of 'religious truths' as beliefs hallowed by the inspiration that they have given to Christians (or to those of other religions) and worthy of reverence on that account. But a doctrine such as the Virgin Birth had no meaning for him at all, and indeed was 'repellent'. He could see no authority in religion, either in the Bible or in the teaching office of the Church or anywhere else. He could make no sense of the belief that 'Jesus died for me'. He could not address Jesus as divine.

Nevill refused to believe that God is omnipotent, in the sense that he breaks natural law or performs miracles to create faith. Ruth and Nevill suffered the pain and grief that one of their two daughters was born handicapped: 'If God were omnipotent, it would be impossible to love him, seeing what he has done to my handicapped daughter, for instance.' Only a few days before he died I had a letter from him, telling me that he liked something I had written in a book: 'We do not know the options open to God.' He wrote:

Exactly; we may guess that the logically impossible is impossible to God. Probably it was logically impossible to prevent micro-organisms developing which cause sickness in animals. It was however possible to choose constants of nature which lead to the formation of stars and planets in which animals can live.

Nevill Mott was first and foremost a brilliant scientist. But he was also a family man who found happiness in his home. He was an educationist who liked to impart and to share his knowledge with others. He was also a man of many parts and wide interests. As well as all this, it must not be forgotten that, unlike some scientists, he was also a man who thought deeply about the meaning of existence and who arrived at a faith which gave meaning to his life.

THE RT REVD HUGH MONTEFIORE worked in Cambridge for eighteen years, including a Lectureship in New Testament. He was a Fellow and Dean of Gonville and Caius College from 1954 to 1963, Bishop of Kingston-on-Thames then Bishop of Birmingham before he retired in 1987.

Searching for Religious Truths

John Polkinghorne

I got to know Nevill Mott in stages. First the name, as the writer of textbooks on quantum mechanics which figured on reading lists. Then the professor, when he returned to Cambridge and the Cavendish. I was a lecturer in the Department of Applied Mathematics and Theoretical Physics, alongside the Cavendish both intellectually and also, in those days, physically. There was a certain degree of friendly rivalry between the two departments and Mott was a formidable neighbour. Tall, craggy, somewhat remote, he was an awe-inspiring figure, having been a leader in the generation of those who had exploited the new quantum mechanics to explain many physical phenomena. I would have been even more impressed had I known that about then, at a fairly advanced age for a theoretical physicist, he was embarking on the work that would lead to his Nobel prize.

Much later I came to know a little, and to value greatly, Mott the man. I had got ordained and was writing a fair amount about how I understood science and religion to relate to each other. Out of the blue came a letter from Mott, now retired, putting some points and questions to me. They centred on miracles, whether it would be worthy of God to do such a singular act as to raise Jesus and leave an empty tomb. It started a conversation that continued, through occasional meetings and one or two further letters, for a little while. My views are pretty traditional on such matters, Mott's less so. We did not reach detailed agreement but I was left with an abiding impression of a man of utter integrity who would neither assent to anything which he did not fully believe nor flinch from the acknowledgement of any conviction that he had reached in his scrupulous way. Mott's Christian faith was, like the man himself, austere, but it was obviously also of central importance to him. Those few letters, in that spidery hand, originated from a remarkable man for whom, whether in physics or in life itself, the question of truth was paramount.

THE REVD JOHN POLKINGHORNE, FRS, is an Anglican priest. He worked for twenty-five years in theoretical elementary particle physics and has recently retired from being President of Queens' College, Cambridge. He is the author of several books on science and religion.

Nevill's Funeral Address

Peter Walker

Others will speak, on another day, of the distinction of Nevill Mott, Companion of Honour, Fellow of the Royal Society, Nobel Laureate, as man of science, Cavendish Professor, Head of House. I think of him this morning, with affection and with deep regard, as the Master who thirty years ago was kind to a visiting preacher invited to this Chapel by Geoffrey Lampe of blessed memory and then, some twenty years later, was to come out to Ely to talk of matters of Christian belief until, his devoted Ruth with him and accompanied by John Sturdy, his College Dean, he came one morning to receive baptism and confirmation at my hands.

When I think of him and of his faithful, and indeed tender, friendship since, and of Nevill now in the nearer presence of God, the biblical words that come to me are from the prophet in the King James version which he loved:

He hath showed thee, O man, what is good; and what doth the Lord require of thee, but to do justly, and to love mercy, and to walk humbly with thy God? (Micah 6.8)

He was a man of certain clear transparencies. He wished, he said with great simplicity, to be at one with those who believe, and to kneel in church with them, and the heart of the Christian faith as he saw it was simple: that God the Creator is one and the same God whom Jesus saw as loving Father. He brought with him, then, into the Christian Church, as I thank God for today, two perceptions which made of one piece his religious profession and his science.

I place first, and see it as central, his sense of the mystery, indeed the miracle, of human consciousness – the precious gift of consciousness *at all*. To stand under the mystery of it – to refuse to close one's mind, or close one's eyes, or take the gift of consciousness for granted – this was for him a proper reverence, and I see here a great scientist's lesson to us all in humility.

With that deep appreciation of the given-ness of things – that everything is what it is and not another thing – went Nevill Mott's deep respect, as a believer, for God's ordering of His creation as He had ordered it.

There come to my mind at this point some words from a devotional address in Corpus Christi College by the Anglican divine W. R. Inge, then Lady Margaret Professor of Divinity in the University and shortly to become Dean of St Paul's, prophetic, I believe, of our century as it would enfold and singularly apt for the Nevill Mott whom we remember today and who was then (January 1911) a boy of five. Inge's theme, pitched on the temptations of Jesus, was the patience of God: 'Our world is still in the making, and we are in the making too.' 'Christ will not conquer the world,' the preacher continued, 'by commanding stones to be made bread. Nor will he, as head of his Church, work miracles. Supernaturalism, thaumaturgy, true or false, is a short cut to success which he will not take. The Christian is to conquer nature only by studying and obeying her . . . the lesson is too obvious to need many words; it is just this: Play no tricks and cut no knots.'*

'Our world is still in the making, and we are in the making too.' The Anglican Church may thank God gratefully today for Nevill Mott – the Nevill Mott, for instance, who gave himself to Pugwash and the application of science for peaceful ends; who sought to do justly, and who loved mercy, and walked humbly with his God, the God always of *truth*. We had much to learn from him, and to his God, faithful Creator and Father of our Lord and Saviour Jesus Christ, we commend him in love and trust today.

A note by the preacher

Looking back over the years of my friendship with Nevill Mott, I would see that first meeting with him in Caius in 1966, when I came to preach in the college chapel (I was then Principal of Westcott House, the Cambridge theological college to which I had gone from serving as Dean of Chapel at Corpus under the Mastership of another Cambridge Nobel laureate in physics, G. P. Thomson) as having a deeper point of significance in it than a Master's noticeable graciousness to a college guest, memorable as that was. I had spoken in my sermon of Bishop George Bell of Chichester, once a very senior friend, the controversial church leader whose prophetic ministry, not to employ too strong a term, has been steadily coming into recognition in the years even since the 1960s; and the Master had followed up the reference at dinner afterwards and spoken of his own respect for Bell's wartime outspoken questioning, in the House of Lords and elsewhere, of the obliteration bombing of the German cities.

I see two marks of Bishop Bell as pertinent to the Anglicanism which commanded Nevill Mott's respect: first, the priority which, as his own Church's foremost worker for Christian unity, he gave, over and above internal issues of 'Faith and Order', to the common witness of the Churches to the urgent human questions of the century. *The Church and Humanity 1939–1946* was thus the title of his collected wartime speeches and addresses. And secondly, his insistence on the primacy, in all things, of respect for truth. 'Nothing

---

* *Speculum Animae*: four devotional addresses given in the chapel of Corpus Christi College, Cambridge, to public school masters and college tutors on 14 and 15 January, 1911 by William Ralph Inge DD, Longmans Green & Co., 1912, pp. 22, 42.

matters so much as the truth.' I take a brief passage from Bell's own preface to a collection of essays by members of the Anglican Communion on *Christian Faith and Communist Faith* under his sponsorship and the editorship of Professor Donald MacKinnon (Macmillan 1952):

> And because of their respect for truth, their careful consideration of the documents and the trends, their sense of justice in matters of controversy, their modesty and penetration, I commend this volume written by these authors as a worthy example of a cooperative effort in the true Anglican spirit.

It is for their *tone* that I quote Bell's words.

I was not to meet Nevill Mott again until, most happily, I found myself next to him at a College Feast in Corpus in 1983 when I was now Bishop of Ely. A remembrance of that earlier occasion, and that it was Geoffrey Lampe, then Ely and later Regius Professor of Divinity in Cambridge and Fellow, and for a time Acting Dean, of Caius, who had invited me to preach, led, from Mott's remark that he was never sure that he had quite grasped Professor Lampe's theology, to a discussion of it which engaged us for the rest of that long feast. I afterwards sent him some of Lampe's writings and my own address at his memorial service in Great St Mary's University Church in October 1980. It was some little time later that he came back to me and, over a period of some months, would come out to Ely and sit in my study, his attaché case on his knees, and open up the questions of belief which (see his mention of this in *Can Scientists Believe?*) had been first brought into sharp theological focus for him when Mervyn Stockwood, later Bishop of Southwark, and then Vicar of Great St Mary's, had asked him in 1965 to speak there on science and belief.

If I emphasise the significance of Geoffrey Lampe at this point – and it was not only Lampe we spoke of, but often, too, of the Roman Catholic continental theologian Professor Hans Küng (for whom Mott had such respect that he said, with his own distinctive touch of humour, that should Küng become Pope he would be tempted to join the Church of Rome himself) – it is because all that Lampe ever wrote is phrased in the same tone of respect for the truth as one finds it and stands under it that I have commented on in Bell. For Geoffrey Lampe, that respect meant an acknowledgement of the element of poetic statement in the central New Testament narratives of God's disclosure of himself (on the Christian understanding) in the birth, life, death and resurrection of Jesus Christ – and in particular the poetry of the Virgin Birth and the Empty Tomb. But that God had distinctively and determinatively disclosed himself in Jesus Christ in love and power to save was quite simply cardinal to Lampe's whole theological position.

Nevill Mott brought to his appreciation of these matters, as I sought so briefly to convey in my brief address in the Chapel of Caius thirty years afterwards, before so movingly they carried their past Master through the courts of the college and the choir sang their *Nunc Dimittis* at the Gate of Honour, the profound respect for God's ordering of his world, including the human frame in its relationship with the physical world, of a physicist himself at the leading edge of the twentieth-century miracle of the growing understanding of its mysteries. His very respect for the Creator whom he so perceived as faithful in the consistency of his creation's order would not allow him to accept any notion of such an intrusion into that order as would seem in plain terms (I do not know that I recall the word in Mott's conversation or his writings, but I believe it would not be too strong a word for him) cavalier. Yet to the profession that in the 'mystery' of Jesus Christ God had disclosed himself distinctively and determinatively he came to feel that he could, as he profoundly desired to do, assent: and kneel in worship with those who could accept as literal truth, as he could not, certain accounts which he himself must interpret figuratively or, as I have expressed it, in the bracket of truth conveyed poetically.

I have a deep respect for such an understanding and retain a profound and affectionate regard for the man who so humbly presented himself to ask that he might kneel in the Church of God and receive, with the devout, strength to follow the Christian way. And, as I tried briefly to convey, I believe that he had much to teach us, and not least in the *tone* of all he said.

DR P. K. WALKER was Bishop of Ely 1977–89. He is an Honorary Doctor of Divinity of the University of Cambridge, where he is an Honorary Fellow of Corpus Christi, St John's and St Edmund's Colleges, as he is of The Queen's College, Oxford, where he read literae humaniores 1938–40 and 1945–47, serving in the Royal Navy in the intervening years. He was previously Canon of Christ Church, Oxford, and Suffragan Bishop of Dorchester.

For Product Safety Concerns and Information please contact our EU
representative GPSR@taylorandfrancis.com
Taylor & Francis Verlag GmbH, Kaufingerstraße 24, 80331 München, Germany

www.ingramcontent.com/pod-product-compliance
Ingram Content Group UK Ltd.
Pitfield, Milton Keynes, MK11 3LW, UK
UKHW021113180425
457613UK00005B/75

* 9 7 8 0 7 4 8 4 0 7 9 0 3 *